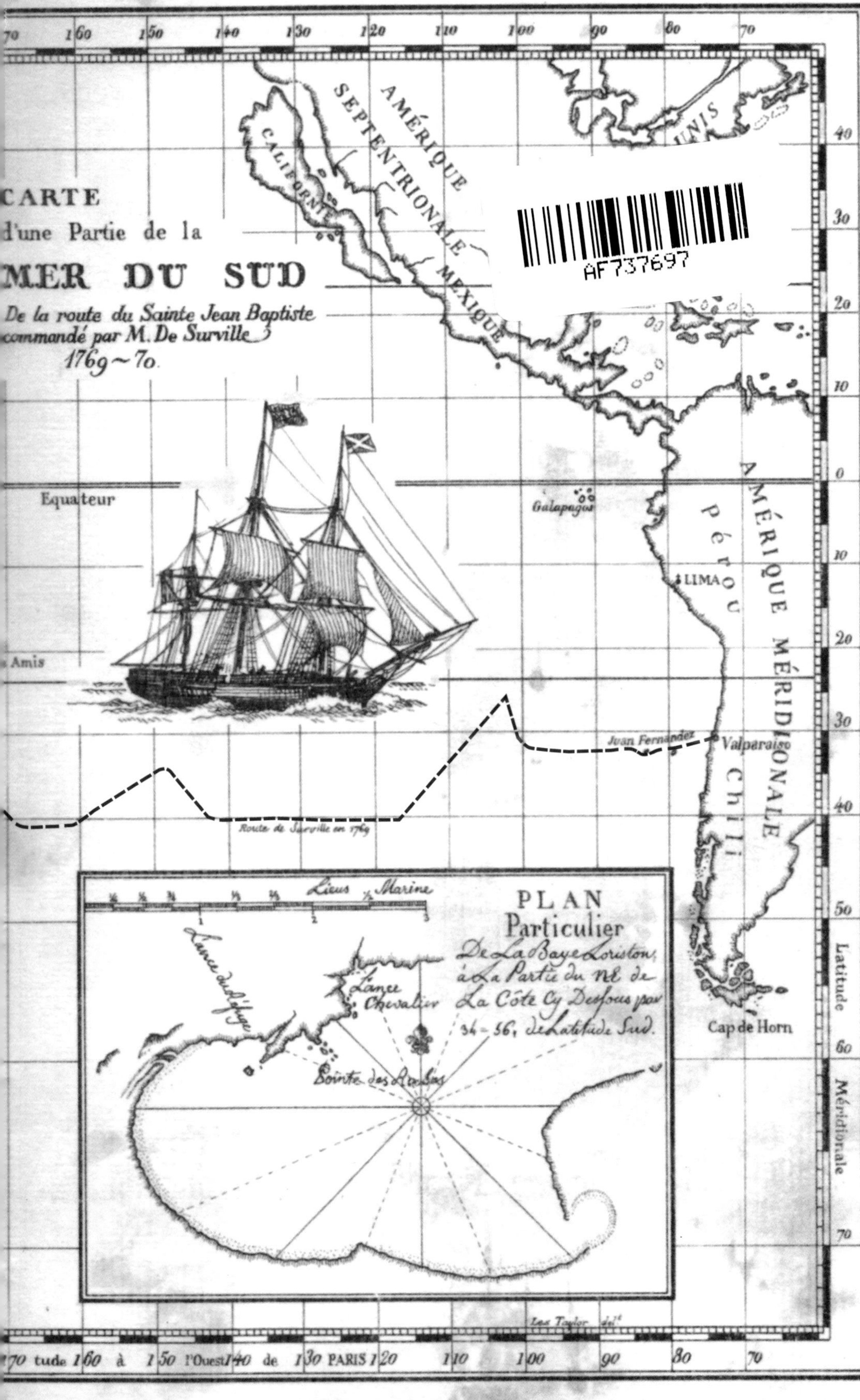
CARTE
d'une Partie de la
MER DU SUD
De la route du Sainte Jean Baptiste
commandé par M. De Surville
1769~70.
AMÉRIQUE SEPTENTRIONALE
CALIFORNIE
MEXIQUE
UNIS
Equateur
Galapagos
AMÉRIQUE MÉRIDIONALE
Pérou
LIMA
Amis
Juan Fernandez
Valparaiso
Chili
Route de Surville en 1769
Cap de Horn
Latitude Méridionale
PLAN
Particulier
De La Baye Loriston, à La Partie du NE de La Côte Cy Dessous par 34-56, de Latitude Sud.
Lieus Marine
Lance Chevalier
Lance du Refuge
Pointe des Rochers
Lon Taylor del
tude 160 à 150 l'Ouest 140 de 130 PARIS 120

THE FATEFUL VOYAGE

OF THE

St Jean Baptiste

Winner of the Sir James Wattie
Book of the Year Award
1970

THE FATEFUL VOYAGE OF THE *St Jean Baptiste*

A true account of
M. de Surville's Expedition
TO NEW ZEALAND
& the unknown South Seas
1769–70

JOHN DUNMORE
CNZM, Officier de la Légion d'Honneur, Officier des Palmes Académiques

HERITAGE PRESS

CONTENTS

FOREWORD
page 7

I. IN FRENCH INDIA
page 9

II. THE VOYAGE BEGINS
page 21

III. MALACCA
page 32

IV. THE BASHI ISLANDS
page 43

V. DISCOVERY
page 53

VI. THE EMPTY SEA
page 67

VII. IN NEW ZEALAND
page 80

VIII. KIDNAPPED
page 91

IX. THE PACIFIC
page 102

X. END OF AN ODYSSEY
page 116

INDEX
page 125

First published in 1969 by
Pegasus Press, Christchurch
This edition published in 2022 by
Heritage Press Limited, Feilding, New Zealand

Design by Nick Turzynski, redinc. Book Design, www.redinc.co.nz

ISBN 978-1-99-116300-4

A catalogue record for this book is available from the National Library of New Zealand.

FOREWORD

AFTER Abel Tasman's visit of 1642, New Zealand remained an enigmatic coastline tentatively drawn on most maps of the Pacific. For one hundred and twenty-seven years no European ship, as far as we can tell, ever ventured near. Then in 1769 two ships appeared on separate and unrelated voyages of discovery. By an odd coincidence, both navigators — James Cook and Jean de Surville — sailed along the same stretch of coast within a day or so of each other without ever meeting. Cook sailed on to fame and de Surville to death and almost complete oblivion.

At times de Surville's voyage received mention in a history or in a book of essays, usually a brief and inaccurate mention, and always in connexion with his appearance in New Zealand waters, so that he seemed to have sailed in one day out of the mist and to have disappeared again as soon as he left the New Zealand shore. Yet records exist, most of them the officers' own journals.

Labé's journal, for instance, is a voluminous one consisting of more than three hundred closely written pages; Pottier de l'Horme's has nearly four hundred, but his writing is much larger; de Surville's log is shorter, with the last entry dated a few days before his death — for he was in the habit of writing up several days together; Monneron is the author of two fairly extensive reports based on Labé's journal and on his own recollections. All these, in fading brown ink on heavy laid paper, sometimes torn and stained, are kept in the National Archives in Paris. In addition there are accounts, some of them tendentious, by eighteenth century geographers, whose interest in the expedition centred on the fact that de Surville rediscovered the Solomon Islands; and a small number of legal documents and letters relating to the voyage.

Although in the main the journals consist of navigational information and routine shipboard matters, they are often vivid enough to bring a scene to life. They certainly reflect the character of their authors, their foibles, their worries, the occasional flash of anger or despair. The present account is based on a reading of all these sources. It is the reconstruction of a voyage on which bravery and fortitude were no match for ill-luck and the baseless fabric of a vision. It tries to introduce to the non-academic reader a little-known episode in the early history of New Zealand and, at the same time, to pay a small tribute to those who suffered and died on that long ago expedition.

PART I
IN FRENCH INDIA

LONG after it was over — when he had returned to France and the drawn-out nightmare had begun to recede into a more bearable past — Guillaume Labé was inclined to accept it as a trick of fate, mean but inescapable, to be endured with all the other misfortunes which man encounters on his way. He himself had survived and returned, impoverished but alive, from a voyage which had taken the lives of seventy-nine of his companions: in retrospect it suggested to him blind destiny, a meaningless impartiality. Those who died, had they been granted the gift of post-mortal speech, might have been less content to describe what had befallen them as mere vicissitude.

It was in January 1769 in the town of Chandernagore, a French settlement north of Calcutta, that destiny, or fate, call it what you will, selected Guillaume Labé. In many respects he was an improbable victim, no man to be tripped easily by sudden mischance. He was a short, square, solid man in his late thirties, a native of St Malo, where the caution of the Norman peasant tempers the adventurousness of the Breton seaman. St Malo, squatly built on a rocky peninsula, looks out to sea on three sides but shelters from its threat of storms and raiders behind thick stone ramparts; it sends its ships to the proved fishing grounds of Newfoundland and to the dockyards of the New World as it once did to the fabled ports of the

Orient, while retaining its links with the farmlands of inner Brittany and of nearby Normandy.

Labé had sailed as a boy to the Indies, both the East and the West. He had fought the English in the frequent wars of his turbulent century, and he had risked his capital in a number of trading ventures; but recklessness had played no part in this. His father and his brothers had all gone to sea, his home town lived by the sea, it was natural that the boy should join one of the many sailing-ships in port and from aloft among the festooning rigging watch the spires of St Malo dissolve in the distant haze.

It may in fact have been a desire for the quiet comforts of home which turned this sensible, rather melancholy man into a potential victim. A very lack of adventurousness sometimes leads a man into taking an unwise risk, because out of the greater profits which danger offers he can sometimes purchase ease and security. Was Labé, on that day in Chandernagore, thinking of a landlubber's life, of the possibility of staying home while others went to sea for him? Who has not dreamt of sudden wealth or, more modestly, of some windfall by which one could enter the charmed circle of financiers and brokers, where inside knowledge is a passport to profit? Thus the very things which should have caused him to turn aside, his native caution, his prosaicness, his desire to leave French India and sail for home, may have conspired to drive him on to rash and heady adventure.

Part of the explanation may also be found in the atmosphere of India itself. Labé lived in Pondicherry, the main French seaport on the east coast of India. It was not merely an exotic city filled with the strangeness of India, it was a city of magic. Speculators, merchants, clerks even, came out from Europe and — unless disease carried them off first — went home within a few years wealthier than they could ever have dreamt. The English had their Robert Clive who at the age of thirty-five had returned to England with a fortune in excess of £300,000; Pondicherry had had François Dupleix, friend of princes, businessman and governor, whom rumour had credited with untold wealth in gold and precious jewels.

AN ENGRAVING OF PONDICHERRY IN THE 18TH CENTURY.

Pondicherry itself, a mass of ruins four years earlier — the English, who had sacked it in 1761, having destroyed every building down to its foundations — had risen like a phoenix, as it had done before in similar circumstances, and the new city, with already thirty thousand inhabitants, was almost complete. Its wide streets were bordered with spacious homes; it had schools, churches, Moslem mosques and Hindu temples. In the narrower twisty streets of the commercial quarter shops and stores bulged with Indian cloth, muslin, silver- and brass-ware spilling out into the roadway, awaiting shipment to France to be exchanged for iron and silver ingots, for heavy French cloth and wine. The surrounding district, its prosperity recovered, provided the city with rice, betel nuts and fruit, as well as poultry, game and bullocks. The Governor's house stood white-painted inside its immaculate gardens and was filled with rich hangings and gilt furniture, a symbol not merely of an earlier magnificence, when under Dupleix the city had so nearly become the centre of a great French empire, but of resilience and optimism. Pondicherry encouraged every dream and, in the person of its present governor, Law de Lauriston, symbolised also a continuing defiance of the English and a belief in a new future. Dream and hope, and fear of England, could be a dangerous mixture.

Chandernagore was less heady. Situated on the right bank of the Hooghly River, twenty miles above Calcutta, it also had known its brief hour of glory under Dupleix, who as its governor had in a span of ten years raised its status from that of a small riverside town to the focal point of France's maritime trade. But in 1757 an English fleet had bombarded it and it had subsequently spent six years in English hands. By the time it was restored to France, Dupleix had died in poverty and the political situation of Bengal had radically altered. Chandernagore was now strangled by Calcutta, which held the river route to the sea, not as before as a mere trade rival in a neutral country, but as the centre of British power in a British protected area. French merchants could still trade, but their success, their very survival, depended on the goodwill of their competitors. Rebuilding had begun, but progress was slow and the defences in particular were inadequate — both because of the shortage of money and because it was unwise to attract attention and excite jealousy.

Labé had been brought to Chandernagore from a small village downriver outside which his ship, *Le Mascarin,* lay at anchor. A sergeant from a regiment of infantry assigned to the protection of the French settlement in Bengal had sought him out in the only European-owned trading post in the village — a none too clean, none too bright messenger of fate. Standing in the dimness of the doorway, with the almost palpable heat and stench of India swirling behind him, he had told Labé that Jean-Baptiste Chevalier, Governor of Chandernagore, wished to see him.

Labé could have refused; he could have told the sergeant of his ill-health which only a few months earlier had forced him to relinquish command of *La Concorde,* of his desire, indeed of his need, to return to France and recover more fully under the cool grey skies of his native Brittany. The sergeant, awaiting his pleasure with placid indifference, in no way interested in Labé or in his destiny, would not have argued and would have returned to Chandernagore alone to present his excuses. But Chevalier was a powerful man and had been a good friend. And curiosity as well as a sense of duty had played their part.

So he had set off for *Le Mascarin,* had packed a few belongings, and taken his leave. Three days later he was in Chandernagore, noting the few changes which had taken place since his last visit, the new prosperous houses which stood as witnesses to the profits which India could still yield even under the most difficult conditions. But nothing approached in splendour the Governor's mansion. It was situated at Sorelty, just outside the city, and had been built by the great Dupleix himself. It was described as 'a great palace, the architectural beauty of which was an object of admiration to all'.

The antechamber, into which Labé was first ushered, was richly hung with embroidered cloth and filled with Indian-style inlaid furniture. In contrast, Chevalier's study was furnished in the French style, so that, leaving the wealth of the East, one stepped into an outpost of Versailles, with elegant mahogany tables, matching chairs and escritoire, paintings in gilt frames and an elaborate clock.

Jean-Baptiste Chevalier, Seigneur de Caunan, Governor of Chandernagore, was a portly man in his forties, much given to entertaining and well practised in courtly speech and manners. His opposite number in Calcutta often came to Sorelty for the week-end, accompanied by members of the Council and any important visitor from England, to fall under the influence of the Frenchman's hospitality and charm and to agree privately after the visit that there was much to be said for allowing their French rivals to retain their precarious foothold in Bengal.

He had irrepressible enthusiasm, endless energy and undeniable patriotism. His enemies said that all he lacked to get his genius recognised was a little luck and a few brains. But he worked tirelessly to defend the interests of France in India, kept in constant touch with the French military adventurers who, scattered all over the subcontinent, intrigued against England on behalf of their maharajah masters. Chevalier still believed, even at this late hour, that it was possible for France to drive the English out of Bengal with the help of the Indian princes who saw their freedom threatened. His patriotism and his optimism were both to play a considerable part in the launching of the *St Jean Baptiste* venture.

There were four other visitors in the Governor's study. Three took their leave almost at once. All three were merchants — young men with good connexions in France. The two Delessert brothers, rather nebulous figures in French India, were established in Chandernagore and belonged to a family of Swiss settlers who had built up an important banking business in Paris. Pierre Monneron was a member of a large family of traders and planters and lived in Pondicherry. The fourth man was Jean-François-Marie de Surville, whom Labé knew well.

When Chevalier had apologised to Labé for having brought him to Chandernagore in such an abrupt manner and had poured them all a glass of Bordeaux wine, the Governor began to explain what was afoot.

He had had news of an important discovery in the South Seas.

Intelligence had come to him and to Law de Lauriston, his counterpart in Pondicherry, from their agent in Cape Town. An English expedition had put in there recently after a voyage across the Pacific. It had discovered an extraordinary island, rich in spices, with a pleasant climate and peaceful inhabitants. The English had been so delightfully welcomed and entertained, the island was so charming in every respect and so promising that they now had only one desire — to return to it at the earliest opportunity and establish a settlement there.

Labé, with his experience and his down-to-earth attitude to life, had objected that most ports of call felt like paradise after a long voyage and that no one should set much store by mere sailors' yarns. But Chevalier had added further details, in particular that the island was inhabited in part by fair-skinned people whom one officer had identified as Jews.

Jews in the South Seas sounded less strange to him than one might have expected for there had been rumours of their presence before in the Pacific, notably in New Guinea. The Jews had been scattered far and wide during their unhappy history and some, as Labé knew, had even settled in India many centuries before. Who was to say that the lost tribes of Israel might not be found somewhere in the Pacific, in latitudes comparable to that of their native land? This information was significant to them because the presence of Jews meant the possibility of trade. Whatever natives may do or produce — and sometimes they are too indolent or too backward to do anything — Jews will trade. To men like Labé, who were in the East to trade and not to settle, native people were of little importance unless their economy was sufficiently advanced to include commerce, but the presence of people who were known as clever and hard-working merchants was a different matter.

Chevalier made another point. The English clearly intended to return to this island and establish a settlement which would control not merely trade with the inhabitants but sea routes as well. They had driven the French from Canada, they had almost driven them out of India. It was imperative that they be prevented from keeping the French out of the

Pacific. So far no European nation had secured a foothold anywhere in the South Seas. Bougainville a few years earlier had endeavoured to secure a base on the Falkland Islands which commanded the Cape Horn route but he had been stopped by the Spaniards whose South American colonies felt threatened. The Dutch held the long Indonesian chain which barred the route from the Indian Ocean. But the English were obviously seeking a central base somewhere. The first expedition had been Byron's in 1765, and now there were rumours of a third expedition.

Labé had thus been called to Chandernagore because Chevalier and his friends were planning an expedition of their own which would forestall the English.

Profits came into the picture, of course. The French India Company was almost bankrupt and it was only a matter of time before it was dissolved and the monopoly of Eastern trade which it had enjoyed for so long was revoked. Men like Law and Chevalier whose duties had included protecting the Company would then be free to trade wholly on their own account, but would have in addition the responsibility of providing work for the sailors who might otherwise find themselves unemployed. Captains like de Surville, after years in the Company's service, were seeking new opportunities so that they might provide for themselves against the uncertainties of the future.

Jean-François-Marie de Surville urged Labé to join him. He was to command the proposed expedition and he needed a strong partner, a man of character whom he knew and could trust. He was fifty-two years of age, tall and strongly built, deliberate in his speech and in his movements, a man who had been in command of others for many years and knew how to make himself obeyed. Like Labé he was of mixed Norman and Breton descent, but a native of Port-Louis, a royal naval port on the Atlantic coast. He had been at sea from his earliest teens and had served for most of his life in ships belonging to the French India Company. Indeed, though Labé could not be sure of it, it was said that relatives of his had held important positions in the Company. The Cross of St Louis which he was wearing on this special occasion was proof of his bravery; he had won it during the Seven Years' War while fighting with the Comte d'Aché's squadron in the Indian Ocean, in a campaign which had cost his brother René his life.

De Surville stressed that the expedition would bring profits as well as glory, because Chevalier and he wanted Labé to join them as a partner. Labé replied that he had only recently recovered from an illness which had forced him to relinquish command of the Company's vessel *La Concorde,* that he was planning to retire or at least to rest ashore for a while. He would not have left Pondicherry on his present voyage if the Council of that town had not urged him — almost ordered him — to assist the *Mascarin's* captain, one Monsieur Aumezat, who was compelled by circumstances to sail an unfamiliar route to the Hooghly River.

However vehemently he assured them that he was in no condition to sail to some unknown island, whatever new arguments he put forward, Chevalier and de Surville continued to insist: the expedition would not sail for at least two months and even then would first go to Pondicherry so that he could visit his home; special arrangements would be made through Monsieur Law to look after Labé's wife; they would anyhow not leave India for five or six months, as the ship would have to be fitted out for the long journey and trade goods of all kinds would have to be purchased.

Labé's resistance weakened. For had not fate singled him out? The curtain had risen on a tragedy that would unwind as deliberately and as inexorably as a classical Greek drama.

Later when he looked back on its beginnings he was unable to single out the turning point in the discussion or to decide exactly what it was that had won him over. The faerie setting of wealth and unreality in which Chevalier lived had played its part, but Labé had also been moved by the personalities of two men.

One of these was de Surville. The presence of the other, Jean Law de Lauriston, was almost as tangible. Chevalier had shown him Law's letter, sent a few weeks earlier from Pondicherry, in which he stressed the importance of sending a ship on a voyage of discovery to the mysterious new island. As he held the thick sheet of notepaper with the nervous slanting writing and the clear, firm signature, Labé had once again felt himself slipping under the influence of the celebrated governor of Pondicherry.

Jean was the nephew of that famous Scotsman, John Law, whose financial schemes had so nearly brought prosperity to France and to French India, but had instead ended in disaster. Genius or fool? Labé, like many others, could never decide which. There was less uncertainty in his mind about the nephew. Jean Law, Baron Lauriston, Knight of the Order of St Louis, Brigadier-general and Governor-in-chief of France's outposts on the Indian continent, was an able businessman and a brave soldier. Arriving in India at the age of twenty-three as a clerk in the employ of the India Company, he rose swiftly through its ranks and was appointed head of the French post at Kasimbazar near Chandernagore. He remained in charge for ten years — until 1757 when the British army captured both Chandernagore and Kasimbazar. When the English had overrun all the French settlements, Law continued to fight them as an officer in the service of Shah Alam, Mogul emperor of Delhi. Taken prisoner by the English, he was sent to France on parole and returned in 1765 with the title of Royal Commissioner for French India. It was largely thanks to him that Pondicherry had so quickly risen from its ruins, and Labé, like most of the French in the city, had not forgotten how much they owed him. True, there were some who were prompt to criticise, claiming that he lacked a broad vision, that as an administrator he could deal adequately with only minor projects, that his wealth grew more rapidly than the city's; but Labé had never shared their views. No one in any position of power can prevent tongues wagging — and success can often be measured by the amount of jealousy it generates. In a parallel situation, Labé knew what it meant to stand on a bridge at night and bear the sole responsibility for a ship and her crew and suffer the agony of making blind decisions watched by a group of silent officers.

De Surville was Law's close friend. Fortunately for him, he spent most of his time at sea and had thus escaped becoming embroiled in the petty intrigues of Pondicherry. Nevertheless he had been for a time deputy-governor of French India, and Labé suspected that if anything should happen to Law the directors would at once appoint him in his stead and the nomination would certainly be confirmed by Paris. De Surville bridged the gap between those who sailed the seas and those who sent them thither, between the captains and the traders: he was welcome on board any French vessel because he was a skilled navigator and a respected commander; he was trusted by the merchants because he was one himself and because of his knowledge of distant and profitable — and mysterious — markets in eastern seas; and he had political connexions which no one could clearly identify or unravel.

Labé felt bewildered, unable to explain to himself why he had agreed. The thought nagged at him, as difficult to ignore as the onset of toothache, that he had been persuaded into a false and dangerous position. For had he not agreed to join the expedition and, what was of greater consequence, to invest money in it? And then had been dismissed with exquisite courtesy but without being told where the expedition would be going!

Chevalier had smiled, displaying a full measure of charm and *bonhomie,* and read the agreement to him. He was to join the *St Jean Baptiste* as second captain and second supercargo, receiving a captain's emoluments at the rate of eighty-two rupees a month with a commission of one and a half per cent on all sales and purchases. In addition to this, the sum of one hundred rupees was to be paid every month to his wife until his safe return to Pondicherry. And finally he was to be brought into the partnership to the extent of 50,000 *livres* and would receive a share of the profit proportional to this investment.

Labé had signed and de Surville, taciturn as ever, had given him his hand. Then Chevalier had called for more wine and for food to celebrate the contract.

So it was that, on the 25th day of January in the year 1769, in

Chandernagore, province of Bengal, Guillaume Labé, after a long and eminently sensible career as a seaman, agreed to invest a quarter of his fortune in an undertaking whose success depended on the correctness of rumours gathered in the grog shops of Cape Town.

How could he guess, as he made his way back downriver to the *Mascarin,* that the voyage he was about to embark on would turn into a nightmare lasting four years and that most of those who were about to sail with him would die on the way? Fate never forewarns her victims. All she allowed Labé on that day was the luxury of second thoughts and the torture of knowing that they had come too late.

PART II
THE VOYAGE BEGINS

In the name of God and the Virgin Mary . . .

Any wise sailor, before embarking on a perilous voyage, will place himself under God's protection. Those who, like Labé, have greater misgivings will seek the help of a saint as well. The difference between Labé and de Surville may be symbolised by the manner in which each opens his journal — de Surville mentions God; Labé, less optimistic, gnawed even at this early stage by doubts and indecision, places himself in the hands of the Virgin Mary as well.

The first danger came from the British in Calcutta. Chevalier knew that Chandernagore was infested with spies in the pay of the British India Company, and that every move he made was reported to Calcutta. Only the utmost secrecy could save the expedition from being delayed by stratagems which the English and their Indian employees would devise. Chevalier had experienced annoyances of this kind before. De Surville himself had been stopped in October by a group of native soldiers who had forced their way on board the *St Jean Baptiste* to search for prohibited goods, and protests were still, unavailingly, being sent to Calcutta.

It was therefore announced that the ship was being fitted for trading in eastern waters: on this occasion she was to go to Canton. The English were unconcerned. Their position in the Chinese port was well-nigh impregnable and their influence was such that they could afford to dismiss any French attempt to establish an effective foothold there.

Unfortunately the need for secrecy was so great that Labé too was left in the dark. He learnt nothing more after his initial meeting with Chevalier and de Surville. The Delessert brothers, a shadowy pair who also invested money in the venture, knew even less; but this was not important. They were acting purely as moneylenders on the security of the ship and part of the cargo. Labé, having agreed to a subordinate position, was forced to let discipline seal his lips. Unable to ask bluntly where the expedition was expected to go, he could only speculate. Confiding to his journal, he wrote that the expedition would 'sail for the Straits of Malacca and the China and Philippine seas, and proceed beyond to make discoveries in the north and the south, to New Guinea and New Zealand. After carrying out our search and our transactions, we will return to the Philippines, China and Batavia, and back to Pondicherry.'

Realising that this plan encompassed almost all the western half of the Pacific, an area full of unexplored blanks and lands whose true position was a constant subject of argument among geographers, he added: 'I am beginning to have a poor opinion of the voyage.'

Meanwhile there was nothing for it but to lay in stores for the voyage. This took him two months. The *St Jean Baptiste* was a three-masted vessel of 650 tons, carrying thirty-six guns. In appearance rather stolid — she was first and foremost a trading vessel whose ample holds were needed for a multitude of goods — she was nevertheless elegant in the eighteenth century manner. Her stern was particularly graceful, with the nine ornamented windows of the great wardroom topped with the seven lights of the quarter-deck cabin, the central one of which opened on to a balustraded gallery. The carvings, while sparse, were skilfully done, and although Labé was soon to have doubts about her solidity, the ship rode at anchor confidently, her paintwork gleaming, and there was a pleasant smell of newness and freshness about her.

What cargo she was to carry presented a problem. How can one decide which types of goods are likely to appeal to men one has never seen? If the natives of the unknown land were still largely uncivilized, then baubles, trinkets, beads and lengths of brightly coloured cloth would attract them. So calico, gold braid, handkerchiefs, knives, bars of iron and cheap trinkets were bought and stored in the hold. But as they might be more sophisticated, with some sort of mercantile society, indeed if they were wealthy and owned gold and precious stones, as everyone hoped, then trade would be more profitable and more successful if the French could offer the kinds of goods which their compatriots in Europe were so eager to buy. And so de Surville and Labé laid in muslin, embroidered cloth, silks, rich cloth from Bengal, scarves from Kasimbazar, carpets from Constantinople, port wine and spices — pepper, cinnamon, cloves, nutmegs. Many of these goods would also find a satisfactory market in the Philippines or China if the islanders rejected them. Chevalier added a few boxes of opium, for which there was such a regular market in the East that it could serve as currency.

All this, including what the expedition would buy in the French settlements in Yanaon, Masulipatnam and Pondicherry, would amount to two and a half million *livres.* If one adds to this the cost of the vessel itself, of food, equipment and wages, the total figure approached the three million. This was Labé's estimate and it made his paltry fifty thousand pale into insignificance. It was no wonder that his partners ignored him. They had told him enough to persuade him to join them, perhaps even too much. They felt under no compulsion to confide in him any further.

How can one estimate this amount in modern terms? The *livre* was indeed a measure of weight, similar in meaning to the English pound, both being related to the Latin word *libra;* but the French pound had depreciated to a far greater extent. By 1768, the rate of exchange was beginning to steady but only an over-optimistic or ill-advised money-changer would have given more than ten English pence for one *livre* —

and most of them would have been reluctant to give even nine. This would place the total cost of the expedition somewhere around £125,000 in eighteenth century currency. This figure may not have been reached because the partners were not able to buy at relatively short notice all the goods they wanted, and it is possible that they — and their creditors — exaggerated the value of the investment. Even so, it is reasonable to claim that the expedition of the *St Jean Baptiste* cost at least ten times the £10,000 which had been spent to fit out James Cook's *Endeavour.*

It is not easy to estimate this investment in modern terms because of differences between the economic structure of the eighteenth century and that of the present day. Most food was cheaper, most goods were dearer, and the pattern of daily life was quite different. One could use Labé's emoluments and the allowance to his wife as a base. These amounted to close on £300 a year. An officer in the merchant navy in a similar position today could expect seven or eight times as much. This would give, in present values, a total investment for the *St Jean Baptiste* of between £700,000 and £1,000,000. The most conservative estimate could not fail to exceed a million New Zealand dollars.

The news, which reached Calcutta in early March, that the *St Jean Baptiste* was about to sail was noted but not even commented upon. The French sailed down the river on 3 March 1769, watched with indifference by the East India Company's agents. No one was perturbed. No one knew that the French were hoping to forestall the English in the central Pacific, to establish a trading base and at the earliest opportunity a permanent settlement.

Had the officials of the English Company suspected these intentions they might have taken steps to delay and harass the *St Jean Baptiste.* They took French competition seriously, more so than the Foreign Office in London, which considered the French in India to be harmless and the return of a few trading posts to France after the Seven Years' War a prudent diplomatic move balanced by gains elsewhere.

The departure of the *St Jean Baptiste* aroused no suspicion, both because de Surville was officially bound for Manila and Canton and because Calcutta had information which was more recent and more detailed than the report from Cape Town which had led Chevalier and Law to support de Surville's scheme.

The Calcutta officials had detailed and confidential information about the discovery of Tahiti by Captain Samuel Wallis in the frigate *Dolphin*. This was what, in a much garbled form, Law's agent in Cape Town had heard about and transmitted. Wallis had reached Tahiti on 18 June 1767 and had named it King George III's Island. Tahiti is a beautiful island, with a warm and moderate climate, especially in June which is early winter, and although not large it is far more imposing than the atolls so frequently encountered in the Pacific. But a full report on the value of Tahiti had to await Wallis's return to England. This occurred in late May 1768, and information began to percolate back to Bengal in early November. The East India Company saw no reason to get excited. King George III's Island was no Eldorado.

But unofficial news had reached India long before this date, from Cape Town, where the *Dolphin* anchored in February. The reports received by the Company's directors in Calcutta and Madras had obviously been written in more sober terms than those sent to Law and Chevalier. It all depended on how far their respective informants trusted their sources. Wallis and his officers had been cautious and knew perfectly well that they could not reveal any information until after the official reports had been handed to the Admiralty in London. The sailors, however, most of whom were in their teens and twenties, were loud in praise of the island, and exaggerated both its beauty and its wealth. From their point of view, the greatest attraction had been the ease with which island maidens displayed their charms and, in exchange for a nail — for iron was non-existent in Tahiti — agreed to share them with their visitors. Had Labé known that the *St Jean Baptiste* expedition had been inspired by brown breasts and nubile thighs, he would have been horrified.

Behind the partners' reasoning, however, lay the myth of the Southern

Continent. For centuries geographers had speculated on the existence of a continent in the southern hemisphere, corresponding to Europe and Asia and helping to balance the earthly globe. Viewed against this possibility, the expedition was a valid gamble. The English, had they known, might have been tempted to hold back de Surville — but not to imitate him, since James Cook's expedition, already on its way to Tahiti, was in fact, *inter alia,* to search for the Southern Continent.

So the *St Jean Baptiste* was able to sail unchallenged down the river on 3 March. The sailors, who were superstitious like all of their kind and seized upon omens as do those whose lives are exposed to ill-understood dangers, talked among themselves about the loss of the ship's dinghy which had become entangled in some cable during the afternoon and had been crushed against the hull; but it was a minor incident, soon forgotten as the vessel sailed into the Bay of Bengal and south towards Yanaon.

Shipboard life soon settled down to routine. Labé, mildly disapproving of de Surville's choice of course, had time to sort his papers and balance his accounts.

Even his position as supercargo was vague. A supercargo is the officer in charge of all commercial transactions involving the cargo. De Surville, as befitted his rank and his larger investment, was the senior supercargo. But the ship's clerk, Pierre Monneron, seemed to enjoy the Captain's confidence to a greater degree. No mere clerk ought to have a more influential position in a ship than the First Officer. Yet Monneron, courteous and efficient though he was, seemed to know a great deal — more, it appeared, than Labé. Had Monneron also invested money in the expedition? Admittedly he was only twenty-two, but he had two older brothers in India, both merchants, and one of them was a member of the Superior Council of Pondicherry and a close associate of Law's. Was Pierre Monneron on board to keep an eye on things on behalf of Law and Chevalier? This seemed very likely to Labé — in which case Monneron knew a great deal more of what was afoot than he did himself. And this, to someone who had been offered and had accepted a partnership, was far from satisfactory.

Labé felt more at home with his second lieutenant, Jean Pottier de l'Horme, an able, quiet seaman on whom he could rely. Less reliable, he felt, was Hughes de Surville, the Captain's nephew. Relatives on board ships can be a source of trouble, although one had to admit that young men need experience. There were two men of Bearn on board, neither of them sailors. They had been born but a few miles from each other, within sight of the Pyrenees and they retained the odd accent and ebullient manner of natives of that distant province. One was Pierre Dulucq, the ship's surgeon, whose skill was as good as could be expected in an age when doctors prescribed either purgatives or bleedings for most illnesses and cauterisation for most wounds. The other was Father Paul de Villefeix, a Dominican, whose presence on board Labé could never quite explain. Few merchant ships carried chaplains, except as passengers. It could be that de Surville, whose family had always been hospitable towards missionaries and whose great-aunt had been an abbess, was anxious to ensure the conversion of the natives of the newly discovered land. Yet the Dominican, who at one moment could be seen plunged in his devotions and the next was cracking a bottle of brandy, appeared to him to be searching more for adventure than for converts.

The run down to Yanaon gave Labé a chance of examining the ship. He did not like what he saw. She seemed to have been hurriedly built and badly put together 'which means that there is too much play. Her masts are not supported, the shrouds are close together and get in the way of the guns on the upper deck. Her deck is badly formed, with no contours, so that the water cannot run down to the scuppers and lies in the middle of the deck. Her upper and lower spirketings continually gape open.'

There was too much play everywhere, which meant an inability to keep out rain and sea-water, and consequently damage to stores and cargo, a damp atmosphere below decks, vermin and the likelihood of disease. This was far from encouraging to anyone setting off on a long voyage into the unknown. And yet Labé had sailed in worse and older ships. He closed his journal and went up on deck. The *St Jean Baptiste* was nearing Yanaon.

The town, such as it was, was built on a river-mouth, sheltered by a sandy bar. Neither the French settlement nor the Dutch trading post a few miles away had recovered from the recent wars. But Labé drew up a list of what he hoped to get there: water and cattle in particular — the only alternative to salt meat before the invention of canning was meat on the hoof, the animals often being kept in pens on deck.

When they had anchored, de Surville told him that he intended to go ashore the next day and call on the French Resident, Jean-Jacques Panon.

In the morning the longboat was manned, a hundred men were lined up on deck, the gunners were at their post. De Surville appeared, dressed in his blue and gold uniform, tall, erect, severe as always. He saluted Labé and climbed down into the longboat. Labé gave a signal as the rowers started to pull away. The crew shouted in unison:

'Long live the King!'

'And again.'

'Long live the King!'

'And again.'

'Long live the King!'

Then the guns began to fire at intervals of one minute. A nine-gun salute. Labé knew his job and had a great respect for the proper formalities. He took care to record the ceremony in his journal, for anyone to see and to approve.

De Surville stayed ashore several days. Panon had been unable to raise enough money on behalf of the syndicate, not because his credit was bad but because English merchants in the district lacked the necessary funds. Trade was almost at a standstill. The Mahrattas were reported to have reached the mountain passes to the north-west and to be preparing to descend on the plains. The British administrator's office had been evacuated, a single sepoy being left in charge merely to raise and lower the flag. The merchants, knowing that if the Mahrattas came down they would at once loot the stores, had transferred most of their goods to the relative security of an offshore island some miles away. This made it impossible for the French to buy the trade goods they were seeking because they could not afford to wait for them to be brought to town.

When de Surville returned to the ship, he handed over the task of supervising the final arrangements to Labé. He had bought fourteen bullocks, small but healthy beasts, and fifty-four kids. They would not go far, but they would provide fresh meat for some weeks, and there would be further opportunities to buy food. And he had succeeded in finding seven hundred and fifty heads of poultry, the small scraggy animals which were all one could get in such a climate. Labé had to make arrangements for them on deck, to provide stalls and cages and especially shelter from the heat of the sun.

He went ashore and supervised the loading of the hens. There was a great deal of shouting as the labourers endeavoured to direct the Indian peasants who had brought the animals to the water's edge and as the ship's boats tried to find a way through the press of native canoes and rafts. Panon, looking anxious, came up and handed over thirty-five bales of cloth, all he had been able to buy.

The scene was repeated a week later in Masulipatnam. The French agent there, Mangin, was not expecting de Surville quite so soon and, again, not all the goods ordered had been delivered. De Surville could not wait until early May, the delivery date, and he was not pleased with the quality of the cloth Mangin had been able to obtain — forty-five bales of it. But Labé bought more bullocks and hens, together with a few sheep, so that he did not consider his time wholly wasted. When Mangin came on board 'with four other French gentlemen', they were given the customary 'Long live the King' chorus and a nine-gun salute.

The next step was Pondicherry. It was to be the final port of call. After that the great adventure would begin. Labé however was growing daily more despondent. The stores were inadequate — he needed more spars, ropes, spare sailcloth. 'Had I not given my word to Ms. Law and

Chevalier, I would leave the ship, for the voyage has begun so badly that I am already weary.' A matter of personal concern was the food. He was prepared for all the unpalatable meals sailors had to put up with on long voyages — wormy biscuit, tough salt meat, meat that went bad after a few days and was hurriedly thrown into the soup or served up with rice. Water became brackish or infected and had to be filtered through muslin to strain out impurities, alive as well as dead. But officers at least expected to have a few extras — an occasional chicken, ham, saveloys, and wine. Without being a gourmet, Labé expected better meals than he was getting. Meals, it should be remembered, were often the only breaks sailors got in the long monotony of daily routine. Was he not right, therefore, with all his experience, to express his disquiet?

'For a voyage of discovery as arduous as this is going to be, they have not even taken in stores of food for the officers, apart from hens and sheep (but I am not counting on them, they are so quickly consumed and nearly all die). We should have … butter, cheese, hams, but there is nothing of all that, not even any mustard or beans. It is pitiful. No dessert. We are served any old how. Those who have wine drink it. And yet the ship is supposed to supply the food!'

Law de Lauriston's reception banished these thoughts, at least temporarily. He entertained the officers in the Governor's palace, showed them the most recent buildings of Pondicherry, the park around his palace where Indian gardeners were busy watering lawns and laying out new flower-beds, and exercised all his charm. He remained closeted for several hours with de Surville, then came out to the others in the main drawing-rooms and displayed the latest gifts he had received from friendly Indian princes — carved ivory statuettes, inlaid brass, finely embroidered coats. He gave a farewell dinner at which he spoke of their successful mission to China. If anything this remark, taken at its face value by some of the younger officers, served to remind Labé how little he knew of the intended route of the expedition. He had hoped that he might glean some more information in Pondicherry, but the pretence of a trading voyage to Canton had been kept up. It seemed churlish to say anything in the gilded atmosphere that surrounded Law. And the food, Labé had to admit, was excellent.

The expedition sailed from Pondicherry at 4 a.m. on 2 June 1769, slowly passing a clump of tamarind trees waving in the southerly breeze on a spit of land by the estuary. Labé took his bearing from them. At midday he could still see them on the horizon; a little while later they had disappeared.

'God's will be done,' wrote Labé in his log.

PART III
MALACCA

In June and July, when the south-westerlies are dependable, merchantmen could cross the northern Indian Ocean without great difficulty. The *St Jean Baptiste* sped comfortably along at an average of four knots. After the first few days the sea moderated and the sky cleared. Gulls wheeled about lazily, perching for a while on the top yards, careful to avoid the sailors who climbed aloft to take in a reef or unbend a sail; others sat quietly on the water, allowing the ship to catch up with them as they bobbed up and down on the waves, and then rose with a flap of white wings and a flurry of sea droplets to skim over the surface or climb effortlessly into the warm blue air.

A week went by, uneventful, almost peaceful. Labé would often go up on deck in the late evening, before turning in, if he was not on duty, and stare at the familiar stars, and most of all at the dark, almost hidden, eastern horizon. The only sounds were the rustling and occasionally the snapping of the sails overhead, the soft creaking of the timbers, the medley of sounds and sighs which make up the breathing of a living ship, to which was added the faint murmur of the men's voices below deck.

One night he was joined by Captain de St Paul, the officer in charge of a detachment of grenadiers who had been added to the ship's company at Pondicherry.

St Paul, with a landman's respect for the sailor's special gifts, asked whether Labé could yet smell land. 'Most of my fellows would not be sorry to set foot on shore for a few hours,' he added.

The soldiers had been seasick as soon as the ship sailed and several still looked miserable on the occasions when duty compelled them to come up on deck to exercise under the amused stares of the sailors.

'They will soon become accustomed to the sea. Some of course are born to it,' said Labé, nodding towards Gabriel Le Roux who was checking the animal pens. The boy had been discovered hidden in the hold on the second day. He was a freckled lad from Labé's own St Malo who had sailed to India on his first voyage, and had been paid off with the rest of the crew. Some fool of a sailor had told him the *St Jean Baptiste* was bound for home. The lad had crept aboard and hidden behind some bales, only to discover that the vessel was eastbound. De Surville had threatened him with twenty lashes, and then added his name to the muster roll. Now Gabriel was the ship's mascot, happy to take his turn at cleaning out the sties or swabbing the decks.

At that moment, a bank of cloud stretched across the moon began to shift and Labé could discern the outline of a coast ahead. The ship came to life as the lookout man called the bearings. Men ran out on deck and crowded at the rail, while de Surville hurried from his cabin.

'The Nicobars, sir,' said Labé. 'Great Nicobar, I'd say.'

'We should not be this far east. Tack and stand off. We must wait until there is more light.'

When daylight came, a few hours later, the land bore north-east, mountainous and covered with vegetation down to the water's edge. Farther west a spit of low land curved back towards the south, thinly dotted with scrub and clumps of bamboo. A fresh westerly breeze rippled the surface of the water and threw up a line of foam against a narrow reef.

De Surville realised that there was no chance of hauling round to the north. The *St Jean Baptiste* was already too far east, he said, and the manoeuvre would cost him more time than he was prepared to waste.

Labé concurred: they did not need supplies as yet, the island had little to offer anyhow and Malacca would provide far more. Almost to himself, the Captain replied that he was not interested in supplies, but in the presence of a Danish settlement. Then he turned to his partner with a look of embarrassment, as if caught off his guard and forced to impart a confidence which he had not intended to share, and explained that he had been asked to verify certain rumours concerning a post which the Danes were believed to have set up on the island. Situated as they were close to the entrance to the Straits of Malacca, the Nicobars could be used in wartime to bar the route to the East.

But the matter was not of sufficient consequence for de Surville to alter course. There was no lack of shipping in those waters and someone else would have to investigate. The ship sailed on and they watched the heights of Great Nicobar gradually lose definition and turn into a broad grey line which slid little by little into the horizon. Labé was left wondering what other secret instructions Law had given de Surville and how many he himself would be allowed to discover.

The next day the *St Jean Baptiste* was preparing to enter the Straits of Malacca and this required skill and patience in the days of sailing ships. The winds are irregular, the tides strong. All that a merchantman under sail could do was to zigzag her way south, making at one moment for the coast of Sumatra, at the next for the Malayan peninsula and anchoring until the tide turned. The *St Jean Baptiste* was unlucky. Calms succeeded unstable winds. The tides ran fiercely against her when the wind dropped. On some days the ship's run did not exceed twelve miles. It took nineteen days to cover the seven hundred miles from the Nicobars to Malacca, a painful navigation requiring repeated sounding and frequent manoeuvring, with three men constantly on duty heaving the lead and calling out the depths, one group standing by ready to climb aloft, another to lower the anchor, and extra men on lookout duty. Every day the temperature rose and the air below decks became more unbreathable.

The strait had a bad reputation. Sumatra, no one would land on, but the Malayan coast was equally dangerous. Labé went ashore on the small island of Pulo Varella to get fresh drinking water and some firewood for the galley. The island was usually uninhabited, but Labé met a group of Malays who told him they had come from Malacca and were looking for turtle eggs. While the water casks were being rolled ashore and filled from a little stream, he talked to them and received a gift of a hundred eggs. If they caught some turtles during the night, would Labé like some? He would indeed. He was searching his bag for some trinkets to give them when one of the sailors drew his attention to the *St Jean Baptiste:* the ship was flying her ensign at half-mast, a signal which meant: 'Danger. Return at once.'

De Surville had recalled his First Officer because one of St Paul's soldiers, who knew these parts, had warned him that the French landing party was running the risk of being set upon by the Malays. Only eighteen months earlier, he reported, a Portuguese ship had been attacked and captured off Pulo Varella. 'I was not prepared to let you run such a risk,' the Captain told Labé. 'Our water supplies are not seriously low.' The expedition raised anchor and sailed on to Malacca.

'Malacca Bay is a fine bay,' wrote Labé when the French arrived there on 29 June. The town itself was less impressive. A fort surrounded by a few unpainted houses stood squarely at the sea edge, while merchants' homes and warehouses were strung along the shore and along the river-bank where, insensibly, they gave way to the Malay and Chinese quarters. It was, however, an old settlement. The Portuguese had first landed there in 1508 and had ruled it for a hundred and thirty years, during which time it had been the king-pin of their Eastern possessions. It was from here that d'Albuquerque had sent an expedition to the Moluccas; it was here that St Francis Xavier had set up his first Malayan mission station. But all this was in a dim past. Portugal's colonial glory had faded. The Dutch had driven out their rivals from this stronghold in 1641, and it was therefore to a Dutch governor that de Surville paid his first official call.

'The Governor took me at once in his coach to dine at his country home,' he told Labé. 'And in the evening a carriage brought us back.' De Surville was well satisfied, but Labé, who had to remain on board, was less pleased, especially when he learnt that Monneron had accompanied de Surville in a second coach. Why should the ship's clerk be so favourably treated? Because he had powerful friends? Labé had not been invited because the duties of a first officer are too important for him to waste time on social visits; they became even more so when it was discovered that the rudder had split. The Governor offered de Surville the use of the port's repair yards, but there was no suitable timber. 'The port's master carpenter dug up for me the keel of a vessel which had lain abandoned for years in a Moslem shipowner's garden.' The only drawback was the cost: the owner knew that the French had no choice and that a good profit was to be made. De Surville had to resign himself to paying three times its real worth.

For Labé the problem of supplies remained paramount. Everything was scarce, prices were exorbitant, it was hard to tell which were the greater rogues, the Dutch or the natives. He bought a few hens, some buffaloes and a few sacks of rice.

'May God grant us good health,' he wrote. 'We will be in a pitiful state otherwise, with no little comforts to make life easier. Should scurvy attack the men, we will lose nearly all of them.'

His pessimism was not shared by de Surville — it seldom was — but hopes of further supplies were dashed when the Governor's manner suddenly cooled. He had been delighted to entertain the French; he had found them pleasant company and had asked them to call on him again. Then all at once he became stiff and formal.

'I believe,' wrote de Surville, 'that if the five measures of rice he gave me were not already on board, he would keep them back from me.'

The change followed the arrival of a ship from Bombay. The India Company had at last heard rumours about the true destination of the *St Jean Baptiste.* What this was the English were not sure, but they did not like the sound of it. The captain of the vessel, the *Argus,* promptly went to see the Governor and although he 'could not know how things really stand since no one knows anything', the Governor sent for de Surville and told him it had come to his notice that the French were bound for 'questionable places'. To the Dutch this meant only one thing — the East Indies, where their own monopoly was rigidly enforced and no foreign ships were allowed. De Surville and Labé realised that the India Company was hoping to arouse the suspicions of their Dutch host so that he might attempt to hold them back. He could, for instance, require them to wait while he asked for instructions from the Governor-General in Java — in which case months would elapse. It was a dark English plot. Was there not among the passengers a Portuguese, a certain 'Mr Caravaillo who has the soul of an Englishman and is as jealous as they are of anything we may attempt'? These Carvalhos were Madras merchants who had become identified with British interests. De Surville decided to take no chances. He weighed anchor at the earliest opportunity and resumed his southward struggle against winds and tides.

The expedition nearly came to an end at this point. One afternoon Labé rushed out of his cabin to investigate a commotion below decks. Sailors were running everywhere shouting that a fire had broken out in the hold. But when he reached the source of the disturbance it was all over. Two sailors were dragging a smouldering mattress which under the supervision of the boatswain they just managed to drop overboard, while in a corner a young man hovered about anxiously.

What had happened?

Fire at sea in the days of wooden ships was a danger which every sailor feared. The strictest orders were issued to prevent an outbreak below decks where cramped conditions, packed stores, hammocks and tarred ropes increased the risk; and generally the sailors obeyed them scrupulously, knowing that their own survival was dependent on constant watchfulness. Labé saw the boatswain berating the cowering youth with rough eloquence, yet when Labé approached the man chased him away and waited until the lad was well out of sight before answering questions. Labé felt suspicious.

Spiriting a guilty man from an officer's presence was an old trick. He had done it himself. It enabled the man to avoid the effects of immediate wrath — sending for him later for admonition or reprimand was sometimes too much bother and by then, anyhow, anger had had time to cool. The men, he well knew, always sided with one another, even though they might set their own penalty for the one who had endangered their safety.

The boatswain turned to Labé and explained that the lad, a young steward, had been checking the stocks of brandy. Unable to see in the gloom and wanting to assess how full a barrel was, he had lowered a lighted candle over the brandy. He had allowed the flame to come in contact with the spirits, which had caught alight. He had burnt his hand slightly and called out in pain and fear, sending a surge of panic through the ship. Fortunately the boatswain was nearby and, running up with a couple of men, had used a mattress to smother the flames.

Labé, determined to protect the safety of all from the stupidity of the few, sent the boatswain in search of the steward. The man had done his best to help him escape a lashing from the First Officer's tongue or, more likely, from the end of a rope, but the matter was too serious to be overlooked, and Labé could be a stern disciplinarian when the occasion demanded it.

The expedition continued on its way south, entering the narrow part of the straits, struggling on towards the labyrinth of channels and islets at the tip of the peninsula where one day the city state of Singapore would rise, but which was then deserted, except for an untidy fishing village where no ships' captains ever thought of calling.

On 23 July de Surville dropped anchor off Pulo Tioman, a large island close to the southern coast of Pahang. There were a few huts in the bay, homes of Malay fishermen. As soon as he had anchored he lowered two boats and sent a watering party ashore with a detachment of ten soldiers. 'The local people, who saw we were well armed, were most welcoming.' As Labé knew, this was a country where strength ensured safety and where hospitality was ever uncertain.

But Pulo Tioman still proved costly. Water was easy to obtain and the casks were all filled in less than forty-eight hours, but, as the sailors began to heave at the capstan, the cable, which had been rubbing against a coral shelf, snapped. There was no hope of retrieving it. The loss of an anchor was always a serious matter to a sailing ship — manoeuvring against wind and tide in a narrow channel could not be done without anchors, as was evident from the *St Jean Baptiste*'s struggles in Malacca Strait, and many a ship has been lost, drifting helplessly in deep water towards a steep coast or a coral reef.

It was at Pulo Tioman that Labé for the first time recorded a desertion in his journal — the first of many. The deserter was a Malay from the Dutch East Indies whom de Surville had been using as an interpreter in his dealings with the fishermen of Pulo Tioman. As the last boat was preparing to return to the ship, he turned and fled into the forest. De Surville wasted no more time in trying to recover him than he had with the anchor.

A few days later he put into Trengannu. It was a large if rather unattractive settlement. 'The town is badly built with rough wooden huts', wrote Labé, 'some with tiled roofs, others with roofs and walls covered with coconut leaves. They are raised on piles eight or nine feet high. You climb into them up a bamboo ladder.' The royal 'palace' was no different, being surrounded by a palisade of heavy planks most of which had fallen down through disrepair. But the Chinese quarter was tidier, consisting of neat houses and shops. And above all, Trengannu was an active commercial centre.

Everything revolved around the Sultan or — since he was away on a military expedition — his representatives. No one was allowed to trade with foreigners unless it was in the Sultan's name; merchant ships were sent by him to Siam, Cambodia or Java; and no one dared to break this monopoly. For Labé, who was now able for the first time on the voyage — and, though he did not know it, for the last time — to record a complete set of commercial transactions in his journal and calculate the commission he would earn on them, the trade arrangements at Trengannu were simple and efficient. 'You land with your boats; you are taken to the king's merchant a short distance from the landing stage. From then on you have merely to state your wants and display your wares' — and, he might have added, haggle courteously, not forgetting to present, with discretion and civility, a gift or two to the Sultan's agent who, in serving his royal master, has many cares and expenses of his own.

Everything went smoothly. There was a good market for opium, a lesser one for knives — the ship had a large supply of Flemish knives, cheap but of poor quality, and the Malays, unhappily for the French, were connoisseurs in knives. There was certainly a ready market for muskets. The Sultan greatly needed muskets. Was he not at this very moment at war with a troublesome vassal who had fallen behind with his tribute? And there was also the worrisome question of Malacca: the Sultan, the French were told, 'was also king of Malacca'. Unfortunately the Dutch were in possession of that town and showed no sign of giving it up or even of acknowledging the Sultan's suzerainty. Possibly some other European nation might be willing to assist the Sultan in asserting his ancestral rights. The French, possibly? De Surville was doubtful. Well, French muskets would be welcome anyhow, and without doubt the Sultan, who was fond of foreigners and of trade, would always be delighted to see French merchantmen drop anchor in the river-mouth.

Labé sold four chests of opium and bought food supplies of all kinds: pineapples, pumpkins, cucumbers, greens, eggs, hens 'large and small', bullocks and buffaloes. The latter set Labé grumbling. De Surville preferred buffaloes to bulls because, weight for weight, they were cheaper. But buffaloes are less resistant. 'The bullocks on the contrary keep well on board; but he wanted to have his own way.' Labé, his not to reason why, got on with the work and kept his comments to himself, consigning them only to his private journal.

A reception given by the chief merchant provided a sidelight on Trengannu justice. The French were dining at his house when a strange procession went past. 'A young Malay had been arrested for robbing and murdering a child. He was at once sentenced to death and taken by boat, with his arms tied, all along the shore. The law people stopped from time to time, beating four or five times on a piece of copper, and everybody bowed down before the double lance which had been set up in the boat. The officer of the law called out that this man had been sentenced to death for murder and theft, that the first Malay or European who committed such a crime would similarly be punished by death with no hope of mercy. Such justice shows that the king has established a good administration in his state so that foreigners need have no worries here. The criminal was taken to an island a short league upriver and the affair was concluded at 12.30 in public by means of this double-pronged lance.'

Pottier de l'Horme, that quiet and sound officer whose personality is seldom allowed to intrude in his comments, was as usual more precise than Labé. In Pottier we have a conscientious observer who not infrequently embellishes his journal with an explanatory sketch; this was one such occasion and the marginal drawing shows an implement not unlike a two-pronged garden fork. It was known, he adds, as the 'royal iron', and the execution was carried out by driving this into the condemned man's stomach. The motive of the crime, he tells us, was jealousy. The victim was a youth who enjoyed the favours of a woman the criminal wanted for himself. Having been unwise enough to steal the dead man's clothes he was quickly identified and sentenced.

On the last day of July 1769 the French made ready to sail. They hoisted in the four boats, stored the last water cask, bought their final supplies of fruit and vegetables.

They had kept to a normal course for a vessel bound to Canton. Most of the crew — those who speculated about their future and were not content merely to let each day bring what it may — believed that what lay ahead was a stay in the strange and dangerous Orient and after that as prompt a return as could be hoped for to Pondicherry and Chandernagore.

But for de Surville and Labé the future was fraught with greater possibilities and greater dangers. 'Sailed at 8 p.m. for the China Seas to continue our voyage,' wrote Labé. 'May God lead us safe into port.' De Surville made no special comment in his journal. The ship under his command had as rich a cargo within her as any he had sailed in, the instructions were as charged with hopes and as void of precise details as any ever given to him. He alone knew how much uncertainty lay ahead, what daydreamings and visions of strange new shores and new faces were driving him on. He stared ahead, praying not merely for a safe port for his ship and her crew, but for a settlement like this one, where a wise native king dispensed justice and promoted trade, welcoming foreigners in search of gold and silver, and rare cloth richly embroidered with wondrous patterns no European eyes had ever seen. Somewhere behind the barrier of the Philippines, far to the east, where the sun rose over an unknown Pacific, there was such a land, hinted at by geographers, glimpsed by navigators, yet tantalisingly distant and evanescent.

Or else there was only a mirage, born of the heat and clouds, fashioning strange shapes along the horizon, concealing the gaunt hand of death beckoning him on.

PART IV
THE BASHI ISLANDS

Pulo Condore is a large, lonely island in the South China Sea, due south of the Mekong River delta. Peaceful fishermen and peasants lived frugal lives on it, seldom bothering to cast a glance at the rare sails on the horizon, too concerned with their own existence to spare a thought for strange foreigners on distant ships beating their way up towards Canton.

The *St Jean Baptiste* sailed past Pulo Condore on 6 August at a distance of twenty miles. There was nothing to see, nothing to report, merely an anonymous grey shape, a seamark for navigators. Nothing more.

Soon all that could be seen from the ship was the intense blue sky and the impassive sun. Winds rose and died away, sending the men aloft to take in or shake out a reef. Clouds gathered from behind the horizon, formed themselves into ragged patterns which the superstitious sailors attempted to interpret as good or ill omens; they grew sombre, releasing a sudden squall or presaging a storm. Rain fell suddenly in long silent shafts, then as suddenly stopped, so that the wet decks steamed gently, until the heat returned, implacable.

Day succeeded day. The routine of the ship had long since asserted itself. The rounds of meals, with the food becoming more spartan day by day as the fruit and vegetables ran out and the last buffalo died. The welcome tyranny of the hours, regulating the watches when officers and men silently take over from those whose term of duty has ended. The daily plotting of the course, based on the midday altitude of the sun which — providing the sky was not cloudy — gave a reasonably accurate indication of the ship's latitude. But this did not give the ship's correct position, since the calculations were falsified by the estimated longitude, a matter of inspired guesswork in those days when seagoing chronometers were unknown. Estimating the ship's longitude required a series of operations at regular intervals, in particular the calculation of the ship's speed. For this there was no mechanical device. One man threw the log line over the stern; a wooden block tied at the end floated in the wake, while the line itself ran over the rail; the second man was then required to estimate the speed of the line and hence of the ship. But who could judge how much time elapsed between the passing of one of the knots on the line and the next? For, again, there were no reliable timekeepers. The boatswain, that capable, responsible man, might call aloud the seconds as the rope slid slowly past, but who could claim that this was an accurate method? The errors, inexorably cumulative, created a wide margin of uncertainty on either side of the track which de Surville and Labé in their cabins were carefully plotting. For what of the currents? Who could estimate their speed merely by staring over the side at occasional floating debris? And how far off course did a ship drift during the dark, bleak night watches?

Each day's entry in the ship's journals reflects the absence of incidents. Thus for Sunday 13th: 'Squalls blew up yesterday afternoon. At 3 we took in the fore topsail, at 4.30 the main topsail and kept the foresail, close-reefed. The weather was fine until midnight, but when the moon set we had squalls and heavy rain. At 5.30 a.m. we hauled in the main topsail and the mizzen topsail. At 6.45 we let out one reef in the topsails and another at 10 o'clock as well as one in the mizzen topsail.' It could be almost any day in any one of the journals.

At about this time de Surville should have altered course to follow the route the expedition was officially making. But he kept steadily on

A FRENCH
INDIAMAN

A model similar to
the *St Jean Baptiste*

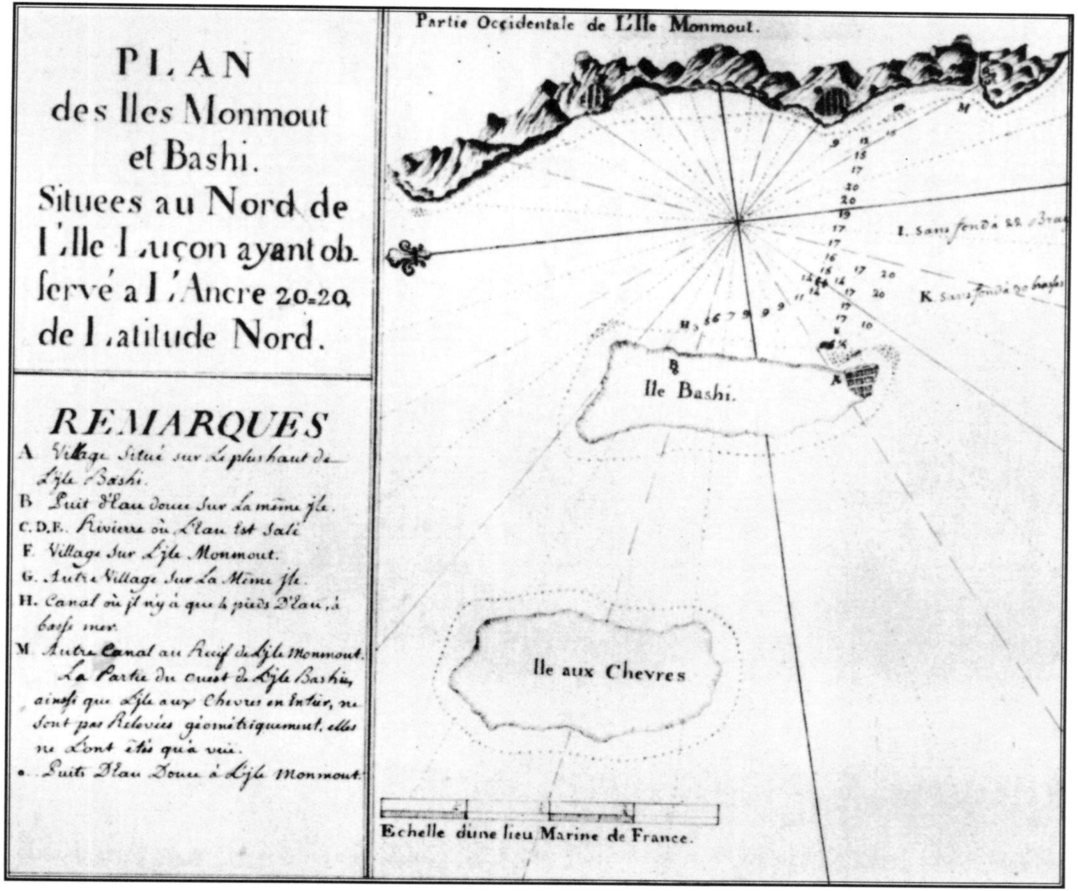

THE BASHI ISLANDS

A map of the Batans probably drawn by Pottier de l'Horme

an East-by-North-East course. The adventure was beginning at last. The officers tried to guess his destination. What thoughts, what fears, assailed the young men bound for the unknown, on a voyage of unpredictable duration? China being ruled out, the possibility of Manila was eliminated a few days later when the *St Jean Baptiste* failed to veer due East. Monneron and Labé, who knew a little, kept their counsel. Others, who were more in the dark, like Pottier and Avice, the young ensign, examined the charts to guess where the ship would go next. Japan was an unlikely choice — it was almost a closed world, except to the Dutch who, under severe restrictions, were allowed to trade there. Formosa and northern China were equally unlikely places for a French merchant vessel. An alternative was the mysterious immensity of the Pacific. It was little known, but it was a scene of increasing European activity and rumours about it abounded.

Then, on 16 August, the north-west coast of Luzon was sighted. Driven by a steady south-westerly, the *St Jean Baptiste* sped north towards Cape Bojeador, the northern extremity of the island. Ahead lay two groups of islands separated by a wide but dangerous channel churned by tidal rips. The expedition sailed past the first group, the Babuyans, and made for the Batans, sixty miles farther north. De Surville started to tack and reduce sail. The men realised that he was going to call at these lonely islands, unfamiliar grey shapes clustering under an alien sky.

About the only thing they knew about them was their name. William Dampier, the English buccaneer-*cum*-genius who had knocked about most of the Pacific, had called them the Bashis after a local drink made from cane sugar. The *bashi* drink, wrote Pottier, gave a feeling of gentle happiness and satisfaction. The men, on being told this, thought it a pleasant name, which compensated for the strangeness of the islands' appearance. It linked them with the grape-growing districts of their own France where provinces bear the names of wines.

But first things first. Water casks were refilled. Pigs, goats, and vegetables were bought: all were plentiful — one of the islands was even called Goat Island — and the natives were both friendly and honest.

De Surville went ashore with Monneron. As soon as he approached, he was surrounded by a crowd of men and women, jostling one another and shouting words of welcome. They were 'round-faced, with a complexion like old olives, and thick, almost straight black hair cut in a circle above the ears'.

'The men,' noted Pottier, in his careful manner, 'do not seem to be wealthy. Many of them have ulcers and scars. One I saw was disfigured by a large cancerous growth on the neck.'

'We cannot escape from this dilemma,' said Labé, 'that to be happy as these simple people are, we must be free from the vices and restrictions of civilization; but those who achieve this freedom must suffer greater discomforts through lack of the material benefits of civilization.'

Philosophical discussions of this kind were not infrequent among the officers, for these were fashionable topics. The Court in Paris seemed corrupt and decadent, and above all incompetent, especially to those, like de Surville and Labé, who had seen the potential of India slip from the grasp of France and the great Canadian colony destroyed. The courtiers of Versailles seemed not unlike the sycophantic nabobs of the great Mogul in Delhi; and the rapacious merchants of the British, Dutch and French East India companies were little better. Everywhere one turned civilization spelt greed, bribery and oppression. By contrast, simple peasants and fishermen, far from the seats of power, led quiet and industrious lives. Was not therefore civilization itself to blame? Did it not bring every evil in its train and destroy the innate goodness of more primitive man? The argument was endless, because those who favoured European culture could point to benefits found only among civilized nations, and to evils and cruelties that were common among savage tribes. De Surville had a higher opinion of natives than Labé who distrusted the theory of the Noble Savage and saw in them much savagery and little nobility; while Pottier, the realist, the scientist, avoided judgment and preferred to record, meticulously but sympathetically, only what he saw.

These speculations, however, were soon ended. On 22 August three men deserted.

No captain liked to lose sailors in this manner. Even if the deserters were troublemakers — as they often were — their successful escape set a bad example to other malcontents who, if they saw no effort made at recapture, might well try the same thing themselves at the next port of call. Thus, although many a captain wrote mentally 'good riddance' in his log after a desertion, he was unlikely to have taken no steps to find the culprits. But catching deserters was an easier matter in well-administered ports where the authorities understood the problem and would co-operate, than in faraway islands where the natives did not even speak a European language.

At first the Captain thought the men might have drunk too much *bashi* ashore and fallen asleep under a tree somewhere; but this theory was soon disproved: a search for their belongings showed that the men had planned to desert, for their spare clothes and the few trinkets they owned had disappeared.

De Surville organised a search party and went ashore. The island was not large and it ought to be easy, he thought, with the natives' help, to find the three men. He walked to the village at the southern end of the island, followed as usual by a crowd of men and women offering fruit and vegetables. Unchallenged by anyone, the French searched all the huts one by one. There was no sign of the deserters.

Enlisting the Bashi Islanders' assistance was relatively simple — at least de Surville thought so as he saw them shouting orders to one another and running off towards the hills. He had pointed to one of the sailors who accompanied him and, holding up three fingers, he had asked them to bring the missing men to his ship.

'I want my three men,' he had said in French. 'Three men like this one. Bring them to the ship. I will reward you.'

But later, as the sailors rowed him back to the *St Jean Baptiste,* de Surville watched a noisy procession marching down from the village, and realised he had failed.

The villagers were bringing him three pigs.

The scene which followed, as one can reconstruct it from the officers' journals, was a mixture of tragedy and farce.

As he neared his ship, de Surville called out to Labé:

'How many islanders are there on board?'

'Two dozen. They've brought us sweet potatoes.'

'Place them all under arrest.'

When he reached the deck he found M. de St Paul's soldiers mounting guard over a group of natives huddled against the wheelhouse.

'Tie them up.'

In the commotion one of the prisoners jumped overboard and started swimming towards a canoe hovering a short distance away. A soldier raised his musket and took aim. De Surville stopped him and called for silence.

Again he tried to make himself understood with a mixture of sign language and simple French. He meant them no harm. He was a good man. *Mapia* was the word they used to indicate goodness and appreciation. Well, he was *mapia.* But three men were missing. They had fled ashore. Once again he pointed to the island and to three sailors. The islanders were to bring the deserters back. Then they too would be *mapia.*

No one replied. The natives looked interested enough — they knew the strange foreign chief wanted them to do something, but they had no idea what. The soldiers looked on impassively and, farther back, the crew watched in silence.

De Surville repeated his message. The silence continued, broken only by the derisive scream of the gulls wheeling over the ship. He made three sailors mime an escape — they crept away stealthily, trying to look guilty, and were caught by a group of lascar seamen who tied them up and brought them before the Captain. De Surville commended the lascars, calling out *mapia* at them several times. The pantomime relieved the minds of the prisoners — the foreigners did not mean them harm after all, they merely wanted an audience for some special ceremonial. They smiled and nodded, and agreed that this was *mapia* indeed.

'They understand now,' said de Surville. 'Untie them but keep six as hostages.'

As soon as they were released, most of the prisoners, although displaying no resentment, jumped overboard and swam to the waiting canoes. Whatever its meaning, whether it had involved danger for them or not, the ceremonial was over and they were tired of it. A few, however, asked whether they could keep the rope with which they had been bound. The French had been kind, and no doubt they would be prepared to reward their audience.

'They understand, you see, M. Labé. They want rope to tie up our men. Let them have it.'

And so the Bashi Islanders, saying *mapia* and bowing their thanks, dived overboard with their lengths of rope.

Hours went by. Canoes came up bearing fruit and pigs. Their occupants pointed to the remaining hostages and it was clear that they were offering gifts in exchange for their release. They offered one pig, two pigs, three pigs per man, since three seemed a magic number on which the white visitors set great store; but they gave no sign that they had understood any of de Surville's sign language. There was no glimmering of guilt, no suggestion of guile: it was plain that the natives were unaware of the presence of the three sailors hidden somewhere on their island.

Finally, de Surville decided to set sail. The deserters would remain in hiding until the ship was well out of sight. Then, possibly, the islanders would understand why the Captain had been so angry, and the three men would pay the penalty for bringing ill-luck to the islands. De Surville released three of the hostages, giving each a length of blue cloth, but kept the last three as replacements for his own men and as an eventual source of information on their country.

'The three unhappy victims of our sailors' desertion were at the portholes of the main cabin. When they saw the tall mountains of their homeland disappear, they wept loudly enough to soften the heart of the cruellest of animals. They were untied at 5 o'clock and each one was given a shirt and a pair of trousers. Their sorrow was then considerably reduced. They began to laugh and now bear their misfortune with the most philosophical fortitude. There are even signs of goodwill showing through and with the kindness which M. de Surville proposes to show towards them I think they will soon forget a misfortune which they could never have expected.'

Thus wrote Pottier. De Surville kept his word. In an age when the press gang was a normal part of life in the main ports and when negro slaves were being shipped under frightful conditions across the Atlantic, he was considerate towards both his men and the native people. The three Mapias, as the sailors called them, were never ill-treated and received in fact more attention than the average seaman. Two were to die of scurvy, but so were many of the crew. De Surville was genuinely interested in them. They came from a country about which little was known and — who could tell? — one day he might be able to bring them home and help to set up a fragile French foothold in a part of the world where only the Spanish and the English, and at times the Portuguese, had so far been able to trade. And in addition — the very voyage of the *St Jean Baptiste* and his own journal prove it — there was in de Surville's make-up a very real and lively scientific curiosity. They would learn French, he would question them about their country, and he would discover through them more than any other traveller had ever been able to find out about their island homes. This was the reasoning which led him to keep these three hostages and to repeat this action in the Solomon Islands and in New Zealand. He was, like a botanist gathering leaves or a geologist chipping rocks, a scientist collecting anthropological specimens. Labé would shrug his shoulders at such nonsense and mutter under his breath when he saw the Captain fussing over them, and posterity would condemn him as a kidnapper and blackbirder; but for de Surville it was the first step in a long journey of discovery.

PART V
DISCOVERY

ROUNDING the north of Luzon was like a leap into the unknown. French charts of the Pacific were few and unreliable. There were no experts to guide map makers; few sailors to provide adequate data. The Spanish had a regular route across the northern Pacific from the Philippines to Mexico and Peru, followed each year by the ponderous Manila galleon, but most of the north was empty ocean. The Dutch had sailed in the south-west down to Tasmania and New Zealand. But much of the central Pacific was inadequately charted. Longitudes were rarely correct, errors of a hundred miles or so being common, and navigators hopefully set sail for islands which were nowhere near their reported position and in some cases were figments of someone's imagination. This situation made the rumoured discovery of a fabulous island quite plausible and even more tantalising. There were too many blanks, too many question-marks.

De Surville set the ship on a south-easterly course and kept his counsel. Labé, as usual, could only guess what his intentions were.

'It seems that M. de Surville wants to keep to a southerly tack to cross the Line so as to make discoveries in that area. I was expecting him to ask me for my views on the course he is following, being partly in charge of the operations with him, but so far he has never given the slightest hint of anything, so that if anything unfortunate ensues, either through the shortage of food and water, or through death and sickness among the crew, I will wash my hands of it all.'

He knew his place and asked no questions. The Captain was in charge and if he chose not to confide in him, Labé — his investment of 50,000 *livres* notwithstanding — could not question him. Nor could he discuss the matter with Monneron or Pottier. The latter, equally conscious of his position, did not even confide his thoughts to his journal as Labé, more senior and in this respect irrepressible, continued to do throughout the voyage.

There was, for instance, the episode of the chickens. Already a number of sailors had fallen sick. Dulucq, the surgeon, asked for chicken broth for them. De Surville and Labé pointed out that there would soon be no chickens left if they were to be included in the crew's daily rations.

'It is not a question of supplying all the crew with chickens, Captain, but the sick. They will not recover if all they are given is biscuit and salt meat.'

'Are you convinced their condition warrants this, Monsieur Dulucq?'

'Completely convinced, sir.'

'Very well. The poultry will be reserved for the sick. And only for the sick.'

'And for the officers' table? It is customary for us to be served chicken fairly regularly.'

'No, M. Labé. On no account. Unless an officer falls ill, of course.'

'This is one of our captain's jokes, I imagine,' Labé wrote in his log. It took him a fortnight to realise that de Surville was in earnest. 'Not a single item of poultry has appeared on our table since that day!' And by that time the growing number of sick made it unlikely chicken would again form part of a healthy officer's meal. The great hope was that a call at some island on the way would relieve the monotony of the diet with fresh fish or coconuts; but although there were islands about none was sighted.

There is an air of failure and frustration about the voyage south.

Each of the officers' logs shows them plotting a course, with erroneous charts and an uncertain longitude, hoping that each dawn would bring a sighting of a telltale shape lying low along the horizon. But day after day the sea remained empty. The expedition was sailing in another world, of endless sea and sky, becoming steadily less bearable as food and water became shorter and the heat greater.

Tuesday, 5 September 1769. According to M. Maurepas' chart which is quite worthless, I am still among the islands of the New Philippines or New Carolines.

Sunday, 10 September 1769. Sailed under the minor sails for fear of finding ourselves trapped among the Pelew Islands, being, according to our reckoning, very close to them.

Thus Labé went on day by day plotting a notional track through a mirage archipelago, inveighing against worthless charts, sudden storms, and impassive cloud banks behind which might be sheltering a green-clad island with a girdle of white coral sand and freshwater streams. De Surville and Pottier, for their part, spent their time seeking and recording in their journals the sailors' traditional signs of land:

Sunday 17th. About 3 o'clock yesterday afternoon a large round log, several pieces of bamboo, and some gulf weed passed alongside.

Friday 22nd. We have continually seen pieces of wood and large whole trees with branches and fruits. During the night when the breeze veered north we had a very pleasant and strong scent of flowers.

They crossed the Line on 23 September. The sea was calm. 'God knows when we shall have wind enough to get us out of these regions,' wrote Labé. 'We have only enough water for 70 days. Our men are beginning to suffer pitifully from the heat and from thirst, and several are affected by scurvy. I do not know what M. de Surville intends to do, since he has never spoken to me of these operations of discovery. I do not know what to think about him, whether it is due to ill-humour on his part.'

De Surville had no idea that Labé was perturbed. He held his First Officer in the highest esteem and never failed to praise him whenever he mentioned his name; but his own proud nature, his faith in himself and his mission precluded him from seeking his officers' advice. What, anyhow, could they tell him? To seek some refuge, some island somewhere to rest his men and refill his water casks? How obvious! And what could they tell him with their sketchy charts that he did not know already? Could their eyes scan the empty horizon better than his own or those of the sunbaked lookouts who hoisted themselves each morning up to the crow's nest? What advice could Labé give him? 'If my opinion had been sought I would have been in favour of starting our discoveries in northern latitudes and sailing among the Carolines and the Philippines where we should certainly have found quantities of food' — in other words, he would have told him to stick to known waters. Was not this the kind of advice Columbus and dozens of explorers after him were given by the pusillanimous? And which only by ignoring could discoveries be made.

The number of sick grew daily. They complained of stomach pains, and their swollen gums were clear symptoms of scurvy. The *St Jean Baptiste* was making little headway: during the last week of September she was almost becalmed, the great sails hanging limply and heavily over the deck. From the 24th to the 30th the total distance covered was eighty miles. The lack of breeze affected the ventilation below decks. The sick were lying in what seemed a dank well of motionless heat, filled with the smell of their sweating bodies and their vomit. Cockroaches bred in the damp penumbra and crept over clothes and food, while down in the hold rats gnawed at the cargo.

October was overcast. It rained in brief sudden showers, bringing at least a trickle of water into the receptacles disposed around the decks under tarpaulins spread out to catch the raindrops. Somewhere to the south, estimated Labé, lay the rocks of Ontong Java or the shores of New Britain — but how far? Fifteen miles, he thought. But the sun set and rose again on an empty sea.

Finally, on the 6th, Pottier wrote down, for the first time as far as he was concerned, a comment on the Captain's intentions:

'It seems that M. de Surville's intention is to reach the latitude of the Solomon Islands so as to call there and refresh our crew.'

The Solomon Islands! The name makes this brief appearance in the journals and disappears. The charts, as usual, were at fault. The Spaniard Alvaro de Mendaña had discovered the islands in 1567 — just over two hundred years before — and he had named them the Islands of Solomon, as a symbol of the wealth he believed would be found within their shores. Then — unaccountably, were it not for the perennial problem of longitudes — they were lost. Were they in the western Pacific, or in the central Pacific, or even just west of the Marquesas? Some geographers believed that their wealth and their size had been exaggerated, or that they did not exist, that perhaps they could be identified with some of the myriads of small islands that are dotted over the Pacific Ocean. Was de Surville's mysterious island, the aim of their voyage, one of the Solomons? Pottier did not allow himself to speculate any further. He was a realist, a matter-of-fact man. De Surville was nearing the latitude where, so some believed, the Solomons were to be found. It was reasonable therefore to hazard a guess that he was making for them, but not reasonable to surmise too much.

When the French finally saw land on 6 October 1769 they called it the Unexpected Island. If Pottier thought that his hunch had been right and that this was indeed one of Solomon's Islands, he did not mention it — which is a pity, since he would have been right.

For Labé, for de Surville, for all the sailors and soldiers on board, land meant at the time less the possibility of great wealth than the likelihood of fresh water and of a break in the seemingly endless sequence of wormy biscuit and salt meat. The cook and his men were alone in knowing how bad the food really was — for they were the ones who went down in the dark noisome stores where rats scurried and cockroaches scuttled; they it was who washed away the maggots beginning to appear on the imperfectly dried meat and soaked the beans to bring out and drown the worms. All that Labé knew — the rest he suspected — was that food had to be rationed, that the water was foul and that the telltale signs of scurvy were unmistakable. Land, be it the Solomons, be it new and unknown or the outpost of the hoped-for continent, meant running streams of clear mountain water, coconuts and tropic fruit, and a chance to buy fresh meat from the natives.

But a land which is found on none of the charts conceals a multitude of dangers. Hidden reefs and treacherous currents can throw a crew to the mercy of savage tribes. A sailing-ship cannot hug an unknown coast too closely, for she depends for her survival on capricious winds. Where a modern vessel can manoeuvre with precision, at half speed or dead slow, or even reverse engines, a ship driven only by sail must tack and veer and trust in the speed of the men up in the yards to furl or unfurl as the need requires. The dangers are even greater when one finds the coast running east and west in a wide bay, sending out its capes and headlands in an arc like embracing arms to imprison the unwary; for here the winds may drop or be deflected, the currents change direction and shoals lie in wait.

Thus, although land stretched ahead, green and inviting, with a high range of forest-clad mountains soaring towards the clouds, de Surville held back and tacked across the bight, his spyglass trained on the coast. In places the coconut trees and the areka palms lined the shore, but in others the coast receded into the haze. Did these breaks indicate the presence of river-mouths or bays, or was he in the presence of offshore islands defending a mainland where the enigmatic mountains rose majestically like some giant's green castle? He spent a week sailing slowly to and fro, making for the open sea each evening, running back to the land with every dawn. Gradually the *St Jean Baptiste* left the bight and progressed towards the east. The coast grew blurred and was lost for a while in another great bight—this was Manning Strait, the wide channel which separates the island of Choiseul from Santa Isabel.

Then, on 13 October — Friday the 13th — the *St Jean Baptiste* hove to outside what seemed an opening in the coast while Labé went ahead to explore in the yawl, under the protection of five soldiers. He saw at first an apparently continuous line of breakers, but he found an opening and entered a harbour, or at least a wide sheltering place formed by a series of offshore islets.

Now, as he signalled his discovery to the ship, he had his first glimpse of the inhabitants of this strange land. They were as black, he reported, as the Kaffirs of southern Africa, their hair was woolly, they wore bones in their ear-lobes and even through their noses, and their bodies and faces were tattooed or painted in rough streaky designs. Two well-carved canoes manoeuvred swiftly in and out of the coral reefs, keeping at a safe distance, the men shouting at one another and paddling away as soon as Labé moved towards them.

When the *St Jean Baptiste* lay safely at anchor in the harbour, which the French named Port Praslin after the Minister of Marine, a minor flotilla of canoes came out and circled slowly round. The natives gestured and yelled, and one was seen to be readying his bow to shoot an arrow at their strange visitors.

'We tried to pacify him by showing him a small white flag tied to a stick,' wrote Labé. But this did not seem effective. There was no hope of communicating otherwise than by signs — none of the Malays, none of the Indians or negroes on board could find any resemblance between the islanders' language and their own, and cries of *mapia* were, not surprisingly, similarly ignored. De Surville showed the man a length of blue cloth, pointed a finger at him, shouted 'For you!' and threw it overboard. Each side watched it flutter down, float for a while on the surface of the water, and slowly sink.

'Bring me another length of cloth and a piece of wood to tie it to,' ordered the Captain. And this time the man was bold enough to paddle up and seize the gift as it floated past him. Some contact had at last been established, and it was possible to ask where water might be found. 'Putting his hand to his mouth as if he was drinking water,' the Papuan pointed to the back of the harbour and invited the French to follow him; but showing him the sun now low on the horizon, the French explained that they would go with him in the morning.

The sick, numbering two dozen, took heart. The scent of flowers and of wet earth filled the ship where for weeks nothing but the smell of sea had seeped through to counteract the dank odours of sweat and vomit. The ship was steady at last, its creaking silent. The peace was broken only by the light lapping of the waves against the wooden hull and the steady tramp of the sentries on deck.

As the moon rose new sounds travelled across the water, sending a shiver of anxiety down the spines of the men. The Papuans had gathered on the shores of the islands and had lit a large fire. They could be seen seated round it, their dark faces and bodies lit by the flickering flames. Their songs and laughter travelled eerily across the black water. Then it was realised that they were mocking the French, repeating what had been said to them and what they were now hearing.

'They replied word for word to what our sailors were saying: when someone whistled on board they whistled back just like our own men.'

'Monkey tricks,' said de Surville. 'They are amusing themselves. It's a good sign.'

'I hope so,' replied Labé. 'It may be a form of challenge. They may be displaying their contempt for us.'

'No matter. The guns are loaded and ready.'

In the morning more gifts were offered: another length of cloth and a bottle of brandy, the Papuans making light of the brandy which they poured into the sea as if it were tainted water, but expressing joy at the gift of the bottle. Some were bold enough to come on board. They presented a fearsome appearance. Their hair and eyebrows were yellow with lime, some had in addition to ear-rings and nose-rings a shell the size of an egg in the centre of their foreheads. Most wore bracelets of human teeth and a narrow loincloth of tree bark. Mistrust was mutual: 'in spite of all the marks of friendship we could think of, they retained their look of mistrust and anxiety and at the slightest movement we made they would hurry back to their boats or jump overboard.'

At 11.30 Labé went off with two boats, seventeen sailors and

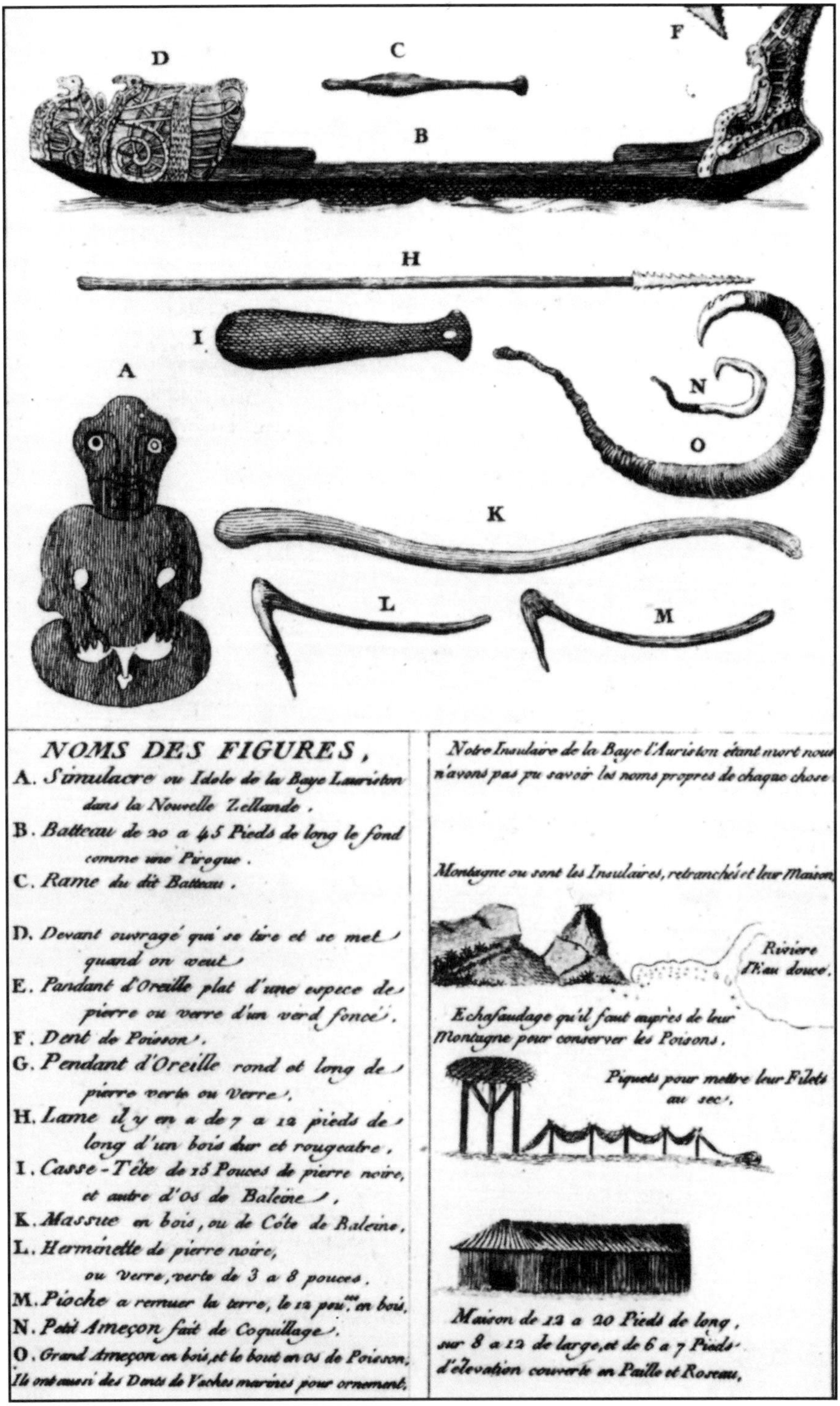

NEW ZEALAND ARTIFACTS

Drawn by de Surville's expedition

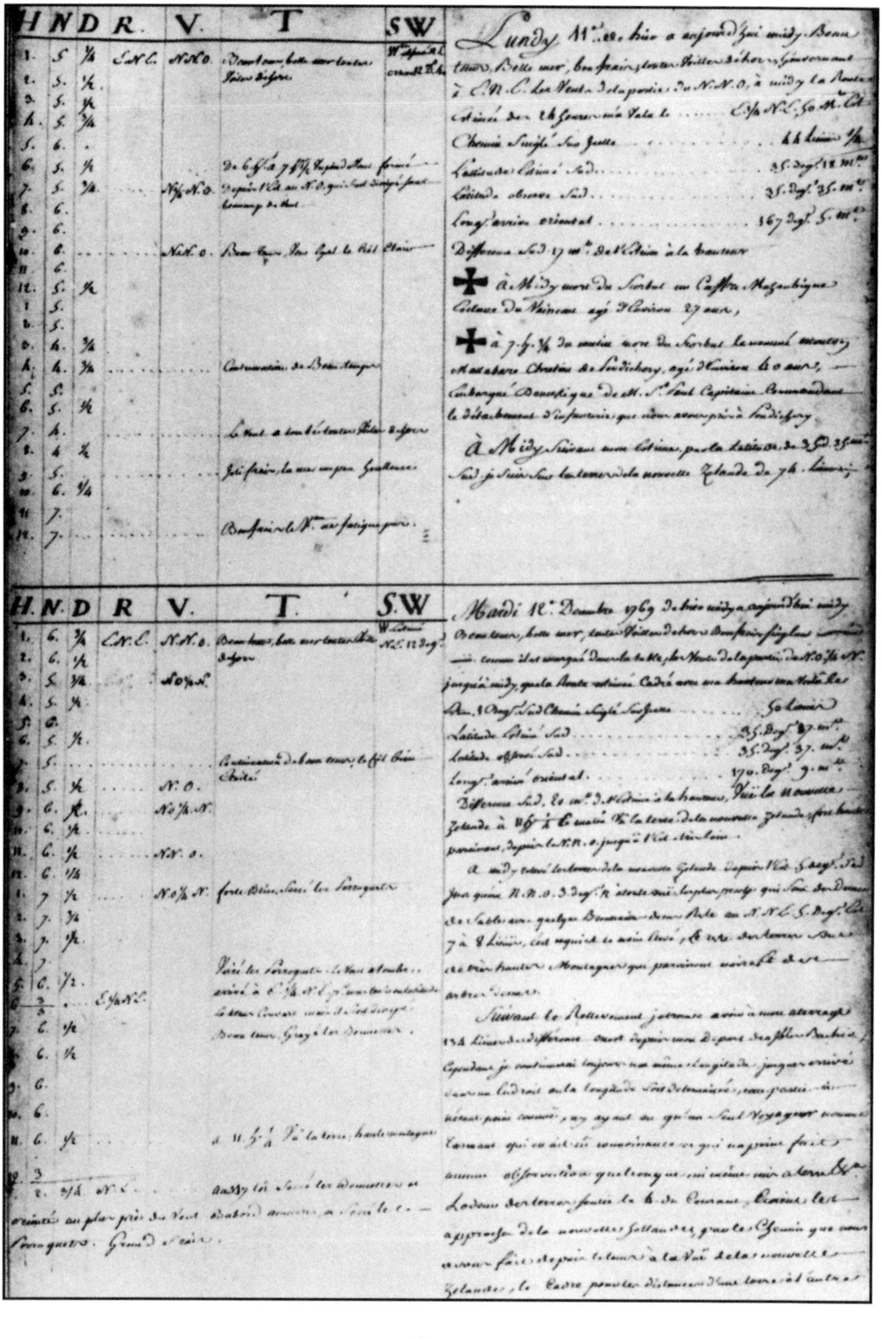

LABÉ'S LOG

The crosses indicate deaths

ten soldiers. De Surville's young nephew was in charge of the second boat. They followed their native friend of the previous day and were accompanied by a large number of canoes large and small. By the time they reached the far end of the bay Labé estimated the crowd at three hundred. But there was no water. The corporal and four soldiers went off to investigate. A little rainwater lay in deep puddles, but there was not enough to fill even one bucket. Labé was able to make his displeasure clear while concealing his anxiety, for he had been led into a narrow and shallow cove where he could easily be attacked. The natives then led the way to another small bay with a narrow strip of coral beach backed by jungle. The French would find water, they were made to understand, if they followed a narrow path some little distance into the forest.

Labé kept his boats away from the shore, holding on to the branches of an overhanging tree to stop himself drifting in towards the beach, and sent half a dozen men in the charge of the sergeant to see what they could find. A quarter of an hour went by. The party returned, having filled two buckets with marshy water.

'We're wasting our time here,' said Labé. 'We'll go back. What's that man doing? '

The sergeant, exhausted and suffering from sunstroke, had collapsed in the water. Labé bent over to help him up. At that moment the natives attacked.

'I saw upwards of two hundred and fifty savages coming out of the forest, armed with areka wood lances seven or eight feet long, bamboo arrows tipped with a very fine piece of bone, and heavy wooden clubs, and others armed with stones. Many of them carried shields to guard themselves from blows. They came at us like tigers. My sergeant was mortally wounded with a lance thrust, a six-inch piece remaining in his body, one soldier was knocked down and fell in the water, another had his scalp split open to the bone, a sailor received a lance thrust four inches deep in the shoulder, and for my part I was struck twice with lances left and right, below the groin, one of the points remaining in the flesh to a depth of more than one and a half inches.'

All this happened in a flash. Labé screamed 'Fire!' and a volley of muskets laid low the first line of assailants. A second volley was fired and this time killed the leader who was standing on the beach encouraging his men and asking his gods to bring him victory. Labé, standing in the bow, received a final blow on the knee from one of the natives who had waded up to the boat; but at this point he had the satisfaction of seeing his attackers turn and flee. A final discharge at the jungle kept them at bay while the boats were turned round and the sailors struck out for the ship.

The battle cost the lives of twenty to thirty islanders and ended de Surville's hope of refreshing his men in Port Praslin. It was now too dangerous to remain there for any length of time; it was impossible to trade or even to obtain coconuts and other fruit from the edge of the jungle where Papuans bent on revenge might lie in ambush. The need for water, however, overrode all other considerations.

Labé now being too ill, de Surville took over.

'I will capture some natives and force them to lead us to a stream,' he said.

Later that afternoon, as the sun began to sink behind the mountains, de Surville was ready to try.

'We sent two of our men from Madagascar in a canoe which M. Labé had brought back. We powdered their Kaffir heads to imitate the local men, circling the ship and making the same gestures we had seen them make. This indeed succeeded. A small canoe with two natives approached and began to examine the ship. Then when I saw them about to leave without coming any closer I ordered our men to fire. One was killed, the other jumped into the sea to swim ashore, but as I had a boat ready he was caught and brought to me.

He is a youth of fourteen or fifteen, dark like the people of Malabar, but no darker. His hair is frizzy, but not woolly; it is coloured red with the lime most of them use, I think on account of vermin. We had a great deal of trouble to seize this young man: he dived whenever the boat got near him. The small canoe with our two Malagasy was forced to go and help with the capture. When he arrived on board he kicked the officer in the stomach and knocked him over. He bit all those who tried to hold him. Finally I had him put in irons.'

The next day the young captive led the French to a spring, not without attempting to escape on the way and biting and hitting his captors. But at least the water casks could be refilled and, cautiously and under the protection of the soldiers, the men could gather firewood and some palm cabbages, while others nailed a notice to a tree: 'Anno 1769. Captain Surville took possession of this harbour in the King's name and called it Port Praslin', and on another: 'Beware of the inhabitants of this place'—notices which the jungle or, more likely, the Papuans watching from the forest, would soon destroy.

Thus did de Surville capture and take away another native. The motive, as at the Batan Islands, was the same — he hoped to learn fully and at leisure what this new land contained and how its people lived. He treated the youth extraordinarily well, having him eat at his own table and finding in him a worthy pupil and in time a friend. Pottier de l'Horme, fair as ever, wrote many months after, when the young Solomon Islander was staying with him in Lima, astonished at all the marvels of European civilization:

'He had received special attention and no doubt his good qualities justified it. Far from taking advantage of it he recognises, on such occasion, that he is treated thus as a favour. His only fault is a slight movement of despair brought about by too great a sensitivity; it is directed only at himself and lasts only a moment. He has a penetrating mind and easily ascertains what people want from him. He would surely learn to read if he were to receive an appropriate education. I have nothing but praise for his probity. He likes dressing up but, from what I have observed on several occasions, giving this up would cost him little. He is fully aware of the value of money, and yet does not prize it as highly as this awareness might lead one to suppose. His main interest is anyhow to satisfy his appetite. … We can assert that he has the happiest disposition and will probably be exempt from many of the failings which are the common lot of most men.'

Labé foresaw a great future for him:

'Being young … he will learn French; then we will be able to get many details on their customs and on what the country produces. … M. de Surville intends to take particular care of him, so as to take him to France, where he can be presented to our Navy Minister.'

But soon after Pottier's note on him in Lima, the youth disappeared from history. We do know that he reached Paris, but there is no mention of his being presented at the court of Louis XV at Versailles. Nor has any chronicler left a mention of his ever meeting France's Minister of Marine — which is a pity, because it would have been the strangest of meetings.

PART VI
THE EMPTY SEA

THE *St Jean Baptiste* sailed from Port Praslin on 22 October. There had been two burials there. The sergeant had died — when Dulucq opened the wound he found the tip of the lance still embedded in his back and had to use pincers to draw it out from between the vertebrae. The second death was that of a seaman from Madagascar, the first victim of scurvy. And on the 23rd two men were buried at sea — another Malagasy and a Breton sailor.

It was imperative to find another harbour somewhere to refresh the men properly, some friendly tribe from whom one could buy food. But the coast looked inhospitable, everywhere reef-fringed, and it was unlikely that other natives in the district would be less savage than the men of Port Praslin.

Unhappily for his men, de Surville's attitude to scurvy was the traditional one. Medical men laid great stress on 'miasmas' and stagnant air. This had the advantage of encouraging cleanliness and the 'perfuming' of living quarters below decks with diluted vinegar, a practice which could not but be beneficial. But scurvy is a deficiency disease due to a lack of vitamin C, the result of a shortage of fresh fruit and vegetables. The disease was common in ships and in prisons, workhouses, and similar institutions where the diet was inadequate; but to the traditionalist the disease was caused by an insufficiency of 'land air'. Since it was unnatural for man to live for long on water or in confinement, the argument went, men unable to breathe normal fresh land air soon fell ill. The connexion between diet and scurvy was in fact recognised, but not the type of food required for good health — de Surville, like most of his contemporaries, blamed the shortage of fresh meat. James Cook's sailors during the voyage of the *Endeavour* were most fortunate because of a treatise by the Scottish physician James Lind, published in 1757, in which advice was given on the use of lime juice and fruits. But little of this new knowledge had percolated through to the French navy. This is why de Surville had done all he could to send his sick ashore at Port Praslin, under strong guard, so they might walk on land and breathe the 'land air'; but he had probably done more harm than good, for they had each time returned soaked to the skin by frequent downpours.

The continent, as the French now called it, stretched on to the south-east — the north coasts of Santa Isabel and Malaita Islands. Days passed without a sign of a break in the reef or an opening in the coast which might herald a haven and a settlement. The condition of the crew worsened. Almost every day Labé, his own wounds throbbing and inflamed, opened his journal and traced the large black Maltese cross which he had chosen as a symbol of death.

Wednesday, 25 October 1769. 2 p.m. Died of scurvy François Cagnard, married, a resident of Lorient, ill these last two months.

Friday, 27 October 1769. During the night, a Malagasy slave aged 28 to 30, died of scurvy. Thrown into the sea two hours later.

Saturday 28th. During the night. Julien Hervé, sailor, from St Malo, died of scurvy. During the morning, two men died of scurvy, one a Moor, the other a Malagasy. …

Sunday, 29 October 1769. Three men died of scurvy this morning. …

De Surville told Labé to have some barrels of port wine brought out of the hold and regular rations served to the crew.

'The men are in desperate straits. They are nearly all attacked by scurvy. We have no fresh food for them'.

De Surville confirmed that he intended to put in at the first suitable bay or port he could find, but the coast so far was too dangerous.

On the 31st, as yet another man died, the *St Jean Baptiste* was sailing towards the island of Ulawa. Smoke rose from several points, indicating the presence of settlements, and canoes could be seen near the shore. The island was well wooded, rising in green peaks above the calm waters, and coconut trees everywhere bent their graceful shapes over the lagoons. The ship slipped smoothly forward with the lightest of breezes, near the entrance of a wide bay. It might be possible to anchor here and come to friendly terms with the natives. Two men heaved the sounding lead at frequent intervals and a lookout who had slowly climbed aloft called out warnings of shoals and hidden reefs.

Canoes approached, with the same type of men as the French had seen at Port Praslin — solidly built, woolly haired, almost completely naked, with the same kind of ornaments — bracelets of human teeth and shells. The sailors waved the customary white flag and threw them gifts. Now some climbed aboard and strode about the deck. One caught hold of the mizzen chains and ran aloft 'like the best sailor I have ever seen'. Then, having examined the ship from his vantage point, he came down to the poop and started to harangue his companions.

The gestures made by the naked savage were passing strange to the prudish eyes of Europeans. So indecent were they that the sailors forgot to laugh and stared open-mouthed.

Monneron, Pottier, Labé, all agreed that they had never seen anything like it. De Surville said tolerantly that traditions differed wherever one went; what was normal among some nations was shocking to others. There were sculptures in Indian temples which were grossly indecent and yet were of religious significance. Labé was less sure of the significance of the gestures they were now watching. Did they imply contempt, suggesting that in the islander's eyes the strangers were as feeble as women and unable to defend themselves against warriors?

The Captain was not concerned. The guns were manned and the soldiers were on duty. There was no danger of a surprise attack as at Port Praslin. The French had already made gifts —bread, lengths of material, and empty bottles. The islanders had seemed pleased and had waved the cloth above their heads in sign of appreciation. Moreover, they clearly belonged to a different tribe from those who had attacked Labé in Port Praslin. They did not wear a bone through their nose, a practice which had given the Port Praslin natives such a fearsome appearance. Instead they wore aromatic leaves thrust through the lobes of their ears. It seemed a more civilized practice.

The native on the poop started to haul down the flag. Two soldiers stepped forward and stopped him. He did not seem put out at being prevented from acquiring this fine piece of cloth and came up to de Surville who presented him with the usual six feet of blue material. Then he seized the Captain in his arms and pressed him against his chest, pointing to the feathers he was wearing and to de Surville's hat. He then stepped back and clasped his hands with a loud clap. De Surville decided that he meant that he too was a chief, and was offering friendship.

'We'll anchor here, M. Labé. You will go ahead of us in the yawl to take soundings and the ship's guns will protect you.'

The bay was now filled with canoes — Labé counted over twenty — moving skilfully over the calm waters, an occasional daring team of paddlers slipping past swiftly under the stern. The island rose ahead of them, green and rocky, with numbers of watchers just discernible lining the distant shore.

The rowers pulled and the yawl began to glide forward, the leadsman calling the depth. Labé had six soldiers with him, muskets at the ready. The canoes began to move away from the ship and circle towards the shore, slowing a little as they came ahead of the yawl.

Suddenly, de Surville, watching the scene from the quarterdeck, saw one of the natives get up in his canoe and aim an arrow at Labé. He called to his gunners to stand by. The man had still not released his arrow and was standing motionless in the swaying canoe when Labé noticed him. De Surville raised his hand, waiting to see what developed before he gave the order to fire.

Labé was not waiting; he fired at the threatening canoe and the battle was on. The attack was unpremeditated — it may not have been meant as such, simply as a challenge or an act of bravado — and it was prevented from developing into a real fight by the two guns which de Surville fired almost at once. A few stones had flown, hurting no one. The Ulawans with a yell had started to paddle towards the stern of the yawl, intending to cut it off from the ship, but the noise of the guns echoing across the bay, coupled with the sudden death of the would-be assailant, struck them with terror and they turned and fled towards the shore.

'Call M. Labé back,' said de Surville sadly. 'This country will only bring us more misery.'

As the *St Jean Baptiste* tacked to reach the open sea, and the men on the yards unfurled the sails, the canoes came back, filled with warriors shouting and hurling stones. The ship fired four shots, dispersing them. Night fell and the silhouette of Ulawa faded into the darkness.

'When shall we be free of this accursed country?' exclaimed de Surville.

He had three more days to wait. When a group of islands hove into sight heralding the end of the archipelago he named them Islands of Deliverance. The farthermost headland his officers named Cape Surville. And then, at last, they were free, but once more sailing over a limitless and uncharted sea. And by then five more men had died.

The course was still south and east. Many years before, somewhere in these waters, the Spaniard Quiros had discovered a pleasant land he had named Austrialia del Espiritu Santo: the Southern Land of the Holy Spirit — the Spaniards were fond of resounding names. There, in what we now call the New Hebrides, were to be found large and safe harbours: the Bay of St Philip and St James, for instance. A south-easterly course should have brought the *St Jean Baptiste* to those shores, or if not, then to the Fijis, admittedly most dangerous, or to the Tongas which Cook was to call the Friendly Islands — and how the poor wretches groaning below decks needed a friendly shore! And if the course had been a little too southerly, then the expedition would have discovered New Caledonia, a long cigar-shaped island laid out almost across their route, although defended by dangerous coral outposts. Here too they might have found, as Cook found, timid and not unfriendly natives.

But fate had decreed that the *St Jean Baptiste* and her crew would search the Coral Sea in vain, that isolated islands would crouch below the horizon or lie concealed behind the heat haze, that the charts would be wrong and the few uncertain coastlines traced by early geographers never materialize.

The *St Jean Baptiste* sailed in a straight line for the New Hebrides until she was within eighty miles of the islands, the erroneous charts then led de Surville to alter course for the south-west and, sailing for undiscovered New Caledonia, speeding in fine weather, all sails unfurled for land and salvation, he was carried north by the currents and sailed past just out of sight of land.

Deaths occurred with tragic regularity — twenty-nine are recorded for the month of November. The ship's routine was altered to cope with sickness and death, and progress made even slower in consequence.

Whereas most captains and officers were accustomed to shout and curse the men to speed them on to the yards a dozen times a day, and to flog them if they were too slow, de Surville began to plan every manoeuvre, to study the horizon for every little sign of a squall, so that his men would not have to unfurl or clew up sails unnecessarily, so that they might never have to work in a rain shower nor be disturbed at night. He watched in anguish as they dragged themselves listlessly across the deck or climbed slowly aloft, resting their foreheads on the ratlines to recover what energy they could. 'When any manoeuvres are necessary,' wrote Pottier, 'one hears terrible cries and moans. The one cries: oh, my arm, another, oh, my stomach, my chest, and so on.'

But these were still active. The rest were below deck, crowded together, stretched out on rough mattresses, suffering from the stifling heat, and rolling about with the movement of the ship. Part of the daily routine was Labé's visit to them.

'I visit the sick daily. It grieves me to see the poor wretches in such a deplorable state. I deprive myself of a number of little extras I took on for myself at Pondicherry; I give it to them, seeing the ship has nothing, such as sugar, syrup, lime juice, coffee, jam, wine, &c.'

We are far away indeed from the stereotype merchant officers of the eighteenth century, loud-mouthed, brutal and inconsiderate. But neither Labé nor any of the other officers could spare enough to save the lives of the fifty or so men below-decks. And so another part of the daily routine required the services of Father de Villefeix.

The chaplain would wait on deck with de Surville and the other officers and a small group of silent men. Four sailors then appeared from the 'tween-decks carrying a long thick plank with on it the body of one of their friends sewn into his own hammock. They would lay the plank on two trestles, the chaplain would step forward, recite the prayers for the dead and bless the body; then each officer and perhaps some particular friend of the deceased would sprinkle holy water over him. The bearers would take up the plank and walk slowly round the deck, circling the foremast, and come to rest by the rail, facing the sea.

'Cut away and release.'

One end of the plank would be raised to shoulder height, and the ropes which tied the body to it would be cut. The body, weighted by two cannon balls sewn inside the foot of the hammock would slide down the plank, plummet to the sea, and be gone. Sometimes a gun or musket would be fired, reminding those below that another of their number had died.

The same ceremony took place every day, sometimes twice a day. There was less formality, fewer prayers, if the dead man was a non-Christian African or Indian; but each time Labé would go to his cabin, open his journal, and trace in the margin the symbolic thick black cross. 'God's will be done,' he added once. He lacked the energy to write more. His wound was still painful; he seldom shaved now, to save water; his face was gaunt and lined; his hair was turning white.

Pottier's journal is a similar catalogue of suffering: 'The men working the ship seem hardly able to stand, a man who asks to be let off from his duties one day finds his legs and feet or his stomach so swollen the next that he cannot rise from his bed, those who are sick soon fall a prey to despair because they see none of their number cured.'

De Surville, alone in his cabin, looked for the hundredth time at his chart. Whatever Quiros had seen the French could not find. His skilled captain, Luis de Torres, had discovered nothing in the regions where the *St Jean Baptiste* was now sailing. There remained Tasman's voyage of 1642. De Surville read the accounts. There were two possibilities. Van Diemen Land, later known as Tasmania, far to the south; but far nearer and with a chance of steady westerly winds, New Zealand.

De Surville called his officers together. At last Labé was being asked his opinion.

'M. de Surville, the captain, consulted us today [24th November] and outlined the state we were in, with many dead and the rest of the crew on the sick list, some attacked by scurvy, and none in a fit state to undertake a long journey. … He is therefore determined to make for New Zealand, discovered by Tasman, the land nearest to us. … He asked for my opinion and I agreed that it was the only course left open to us and that I would have done the same thing had I been in his place.'

Labé estimated that the crew could not last much more than three weeks. Sickness and death are the same, said de Surville, so quickly does the one lead to the other. There was little anyone could do, except pray for a fair wind. Time was lost on occasions because de Surville preferred to sail on a contrary course for a while when the wind changed rather than send the men aloft when it was raining or when he was not sure whether the wind had truly shifted.

Another complication arose. The ship was making water, and it became necessary to work the pumps — twice during each watch — and dry out some of the bales from the hold; this again cut into the rest periods which the surgeon, at his wit's end, prescribed for the crew.

A month had gone by since the French had hailed the Islands of Deliverance and left the infamous Solomons behind. Since then they had had no sight of land. Yet another week went by, during which six more men died, and still the horizon remained empty.

There had been an uneasiness among the officers about their longitude since their departure from the Batans. Tasman's reports might well be unreliable and, if the inevitable errors in their longitude had been cumulative, there was a possibility that they might have missed New Zealand altogether and were now sailing into a void.

According to Labé's own longitude they should have reached New Zealand on 9 December. By the 10th, his reckoning showed him to be well inside New Zealand. But there was no land in sight. Could the expedition have sailed too far north just out of sight of Cape Reinga?

At half past eleven on the morning of the 12th, the man on lookout duty in the crow's nest, listlessly scanning the cloudy horizon, stood up and shouted:

'Land! Land! Land ahead!'

The others were too weak to climb and see for themselves. They stood in the bow staring at the dull sky and the sea, waiting for some green land to emerge. One went below and called the news to the sick. Through the dazing curtain of their fever, they heard the word 'land', and such is human resilience, and so great the curative power of hope, that none died for another two days.

But the land, when it finally assumed shape over the horizon was not green. They saw a long line of sand dunes, a river mouth or bay defended by a long foaming white bar, the grey hulk of a distant mountain. It was, though they knew it not, Hokianga Harbour and the sandy hills of western Northland.

Everyone was now on the quarter-deck. De Surville and Labé, Pottier, Monneron, young de Surville, the surgeon Dulucq who saw hope for his sick whom nothing had been able to save; Father Villefeix, his lips moving in a prayer of thanksgiving. De Surville decided to sail north in search of an anchorage.

The next day a storm blew up. The ship strained heavily in the swollen sea, each lurch and roll inflicting further suffering on the sick. Water ran down into the hold, seeping in from all sides, forcing the men to work the pumps to cope with an inch of water an hour. But at least the *St Jean Baptiste* was making progress, hauling along a sandy coast — part of what is now known as Ninety Mile Beach — and making for the extreme north, for Cape Maria Van Diemen and the islands of Three Kings, where there would be greater chances of finding a harbour.

Fate, however had not finished. She was about to play her slyest, her cruellest and most outrageous trick on the French.

No European had sailed to New Zealand since Tasman one hundred and twenty-seven years earlier; none until despair drove de Surville and his crew to seek refuge in a land about which no one knew anything, except that it was inhabited by fierce tribesmen who had killed four of Tasman's boat crew for no apparent reason.

None that is except James Cook whose *Endeavour* with her well-fed, well-disciplined, scurvy-free crew was at that very moment beating her way up the eastern coast, less than seventy miles away as the crow flies.

The *Endeavour* had on board enough supplies and enough knowledge of scurvy remedies to save the *St Jean Baptiste's* crew. Cook could have directed de Surville to a safe anchorage, warned him about the Maori's bad points, stressed their good points, told him about the country's resources. He might even, if discretion had not held his tongue in check, have given de Surville enough information to solve for him the mystery of the island discovered by Wallis — for he himself had recently been to Tahiti. Certainly, in some way, moved by the humanity of one sailor towards another in peril, he would have helped him.

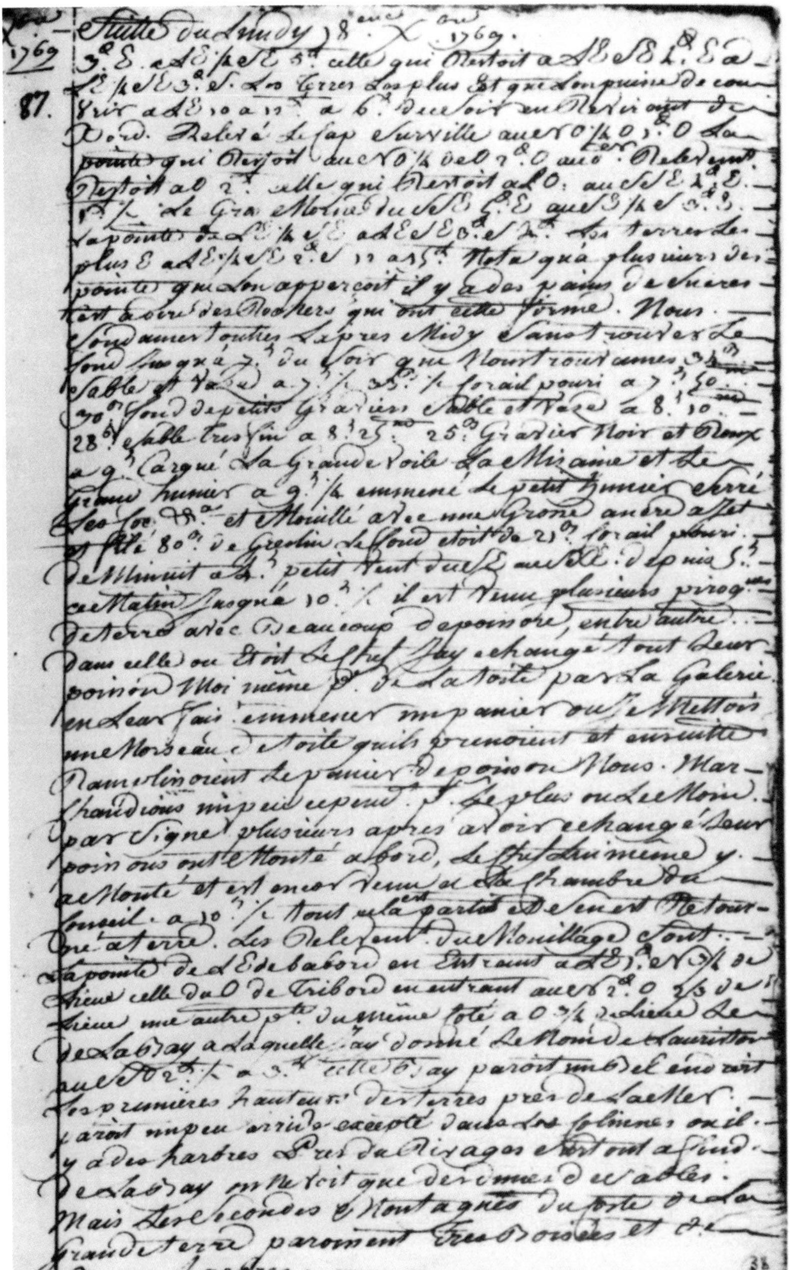
Suitte du Lundy 8. Xbre 1769.

DE SURVILLE'S LOG

The entry records his arrival in New Zealand

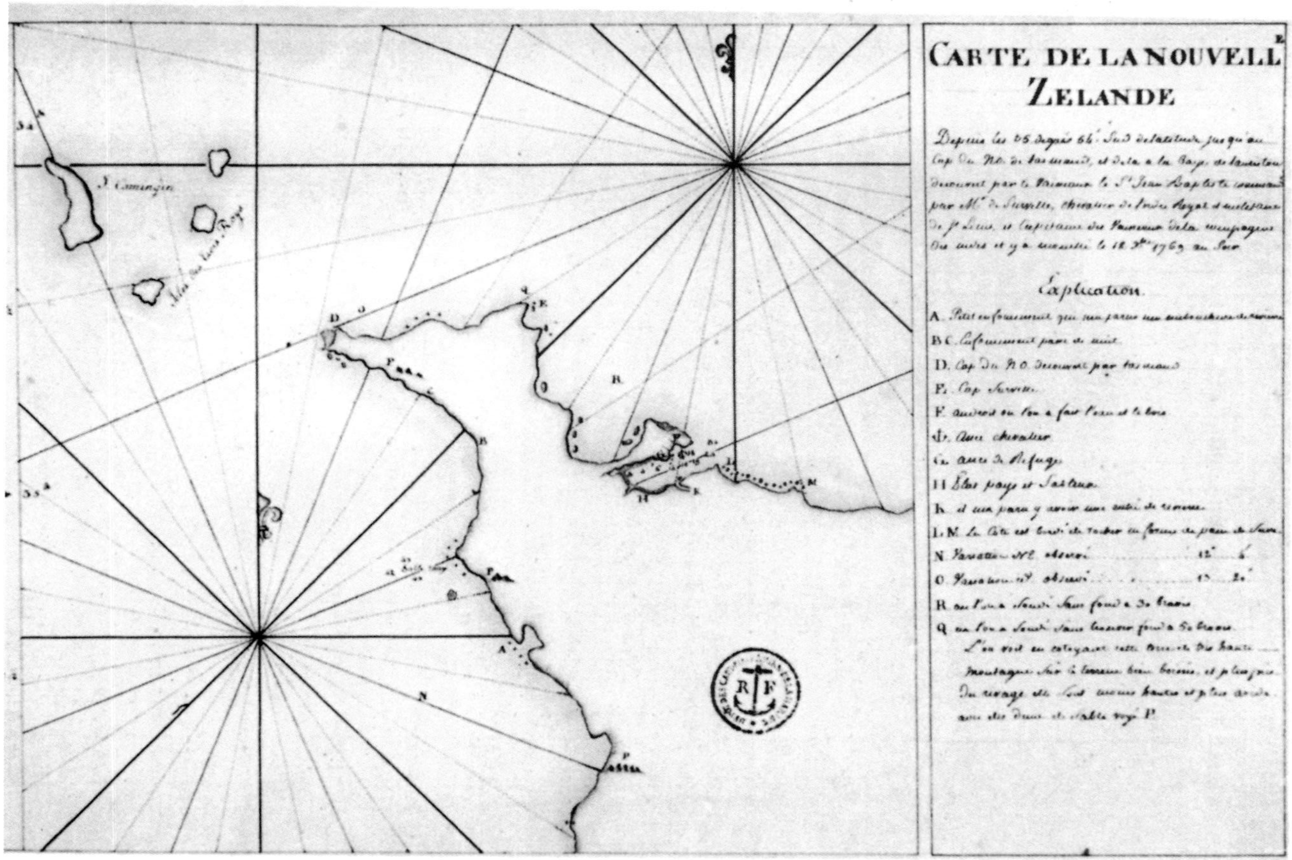

THE FRENCH MAP OF NORTHLAND

But this chance meeting, so crucial, so incredible, and yet now so probable, almost inevitable as the two ships made for the north, could not be allowed to spoil the game.

The storm which sped the *St Jean Baptiste* towards Cape Maria Van Diemen blew the *Endeavour* away from the coast and out of sight of land, holding her there long enough for de Surville to round North Cape and continue on his way south. When Cook managed to struggle back to the coast it was as empty and as desolate as before. All that was permitted was for the Frenchman to be the first discoverer of New Zealand's topmost point, the bleak northern face of Cook's own North Cape, which eventually came to be known as Surville Cliffs.

PART VII
IN NEW ZEALAND

THE Solomon Islanders had looked fierce and been fierce. The Maori, in spite of Tasman's report, looked less frightening and soon proved themselves hospitable to strangers.

The French had their first glimpse of them after passing Surville Cliffs. The wind had abated and the sea was moderate. One canoe, then several, came up to the *St Jean Baptiste,* bearing strongly built, confident looking, brown-skinned men whose hair was tied back in a top-knot. They were muscular, proud in their bearing, and bore large concentric tattoos on their faces and thighs. Above all they came as traders, bringing freshly caught fish to barter with the newcomers.

In the 'tween-decks the sick were told that soon they would have a good meal, that the natives were peaceful, and that before long it would be possible to anchor. In their feverish state, the wretched sailors imagined they could smell the magic 'land air' already rolling down into their stinking darkened quarters. For the first time in six weeks hope had returned, more effective by far than any of the remedies which surgeon Dulucq could bring out of his depleted and inadequate store.

A chief was bold enough to climb aboard. Alone on the upper deck, watched at a distance by a group of gaunt sailors, he stared at the forest of rigging above him, the furled sails, the rising masts, and shook with wonderment. The sailors led him to de Surville who clasped him in his arms. In the main cabin he was offered food and brandy. He spat out the drink, as much in fear as in surprise, but ate the salt meat. Then de Surville gave him a pair of trousers and a vest, both of red cloth, which young de Surville and Monneron helped him put on. In exchange, the chief presented de Surville with a dog-skin cloak.

Thus the news could be sent down to the sick that a friendly exchange of gifts had marked the first contact between French and Maori.

'I think we are saved,' de Surville told Labé. 'If the rest of these people are as peaceful and as civilized as these are, we shall be able to anchor safely and restore the health of our men.'

'May God soon grant us that bounty,' replied Labé, 'for every day we have to spend looking for an anchorage will cost us the life of one man.'

December 17th was a Sunday. Father Villefeix had added a special prayer to his morning mass, and in the evening it was answered. A high bluff was found to be the northern head of a wide bay. The leadsman found ground at thirty-four fathoms, then at thirty, then at twenty-eight.

'Clew up the mainsail,' ordered the Captain.

Labé set about the task of urging on the few men he had left. Next the foresail and main topsail were clewed up, the main foresail was hauled down.

'Twenty-five,' called the leadsman.

'Furl all jibs! The kedge anchor, M. Labé.'

A shrill whistle cut through the deepening gloom. The anchor broke the water with an upspurt of white foam, the coiled cable sprang to life and ran out to follow it.

'Twenty-one. Rotted coral or shells.'

The leadsman straightened up and hauled in his line. The *St Jean Baptiste* had at last come to anchor.

'I will call this bay Lauriston Bay in honour of M. Law, our backer and protector.'

De Surville did not know that Cook had already passed this way and named it Doubtless Bay, or that for the same reason, he was not the first European to sight this part of the coast; but he could lay claim, when he stepped ashore the next day, to being the first European to land on the shore of Doubtless Bay and the first white man the local Maori had ever seen.

They were kind to him. At dawn they came in their elaborately carved canoes loaded with fish. They traded under the stern gallery, the French sending down baskets on a line, lowering a length of red cloth and hauling back the fish placed in the baskets. Trade was pleasingly brisk: although the French were unaware of it, their choice of red cloth was a good one, red being a mark of rank in Maori society. Occasionally a basket would be sent back if the fish were too few or too small and the fisherman would make up the deficiency in order to receive his share of cloth. Soon many of the Maori were wearing their purchases thrown as a light cloak over their brown shoulders; while on board the smell of fish soup, of fish frying, of fish cooked in white wine, spread from the galley to the rest of the ship.

After lunch de Surville ordered a boat lowered, together with the small yawl he had bought in the Batan Islands.

'I embarked a detachment of ten soldiers and gave a sword to each of the rowers, eight in number.... As we approached we could see numbers of natives spread out along the shore and on the nearby heights, as well as in a little fort, some waving a loin cloth from left to right, others a green branch, others a stick. We did not know whether these signs were favourable or not. Finally we reached the shore where the surf was breaking so that we were forced to use the grapnel and to disembark on the sailors' shoulders.'

A chief was there to greet him, his faced marked with the traditional tattooed patterns, deep blue on a brown complexion, a dog-skin cloak over his shoulders, a heavy flat *mere* of greenstone tied to his wrist.

He put his hand out to touch the strangers and 'offered me his nose: it is their manner of kissing'. Labé, watching from the ship, saw de Surville gesturing and pointing to the water casks. The chief took him by the hand, leading him to a small stream running down from the heights. They then sat down on the bank, with the warriors in a semi-circle behind them and the soldiers standing on guard. Dried fish and herbs were brought and set down before them. De Surville ate a little fish and took the greens back to the ship — perhaps he guessed that some kind of vegetables had to be added to their diet if the sick were to recover.

Before he left, the Captain walked up to a clump of trees and touching a few gnarled branches gestured that he wished to cut them down for firewood and take them away. The chief nodded agreement. De Surville was taking no risks — he had no wish to anger the natives by sawing down a sacred tree.

The night was fine. Fires were visible on the heights and along the inlets, signs that the bay supported a large population. The cress and wild celery de Surville had brought back had contributed to a fish-based stew which the men had greatly enjoyed. Was it this, or the land air, or hope — or all three — which made the sick declare that they felt better and one or two come up on deck for the first time for a week and stare at the dark mass of the shore with its pinpoints of flickering fire? The cress and the celery certainly appeared to have amazing effects, bringing a flush to the cheeks and tasting like a heavenly food; but for some it all came too late: a slave from Madagascar died during that night of hope and was consigned to the waters of the bay. Labé traced yet another black cross in his journal, but he would never have to do so again for one of these wretches: 'there are none of them left, all the others are dead.'

The sun rose proudly in a clear blue, early summer sky. The longboat was lowered. Ten empty barrels were hauled down into it; then Monneron, de Surville and his nephew, and a detachment of soldiers and sailors, climbed down the rope ladder. The surgeon Dulucq helped some of the sick into the smaller boat, where he was joined by de St Paul and eight of his soldiers.

The Maori were waiting on the shore as they had done previously, waving branches and weapons. The chief came forward, the rubbing of noses was repeated; but this time he asked the French to wait.

'I agreed to this and he went to discuss with various groups of natives gathered on the neighbouring heights; after which he returned to me and we walked together to the watering place.'

The men rolled out the casks and the woodcutters started work on the trees. The sick disembarked, walking slowly and painfully, helping one another along the beach and towards the long grass at the foot of the slopes, watched by the soldiers on duty and by the Maori above them. At lunchtime de Surville offered the chief some roast pork, the taste of which greatly appealed to him — for he had to be prevented from taking it all. Farther off, seated on the lower slopes, the scurvy cases were eating the same food, with the addition of cress dipped in a vinegar and oil sauce prepared by the cook. Noticing the Frenchmen's predilection for this greenstuff, a few young natives went off collecting it and brought it to them in large handfuls.

But the Maori were beginning to wonder how long their visitors planned to stay and even whether they might not be contemplating some settlement. The ordinary routine of life was disturbed; the food situation might soon become unbalanced if the strangers continued to buy most of the fish they caught. When de Surville went back to the watering place the following day with the same number of soldiers and another ten casks, but this time more sick — since the cress and celery were having a prompt effect and a greater proportion of the men were now strong enough to go ashore — 'the chief came as usual but seemed less friendly'.

He asked the Captain to wait at the shore's edge, as he had done before, and went up to talk to the silent knots of men on the slopes, with yet more silent watchers on the heights above, where the village lay hidden and defended by ditches and stout palisades. When he returned, he made signs that the French were to return to their ship.

'I at once made it clear that I would not. Then taking me in his arms he urged me to wait a while and begging me to draw my sword, he also asked me to give it to him for a moment and told me that he would bring it back. I gladly gave it to him. He took it by the handle and, holding it upright, went back to each group. He returned and offered me his nose, making sign that I could advance to the watering place, which I did, taking the barrels, and set the woodcutters to work and brought the sick ashore.'

The strange ceremony was repeated shortly after this. The chief again borrowed de Surville's sword, this time going up to the *pa* on the hill. The Frenchman agreed — he was still protected by the soldiers' muskets. Meanwhile the economic consequences of the Frenchmen's arrival were becoming more evident: there was marked reluctance on the part of the Maori to sell any more fish. Familiarity was also breeding minor annoyances. There were jostlings, laughter, catcalls, all directed at the sailors and at himself. By lunchtime de Surville was beginning to feel irritated.

'An old man seated near the watering place harangued me ceaselessly in a loud voice, without making one single gesture that might make me understand what he was endeavouring to say. He annoyed me because he would not stop and kept on staring at me and talking.'

One word seemed to emerge time and again from the ceaseless babble. *Atua.* Gods. Was the old priest calling on the gods to protect his people and drive the French away, or was he likening them to gods? Or was he — less politely but equally plausibly — calling them *atua nga tangata:* demons, enemies of man, harbingers of plagues and diseases? Whichever it may have been, de Surville complained to the chief when he came to lunch. Gestures made it clear that the old man was a person of rank, a priest or noble of some kind. De Surville told his servant to take some roast pork to him, then went over himself and tied a strip of red cloth to his lance. 'From that moment he looked happy.'

De Surville was prepared to make any reasonable concession to please the Maori. That is why — even though in time he wearied of it — he was willing to trust the chief with his own sword. He had to refill at least a hundred water casks from the slow-running stream. And it was far easier to buy fish than to send his men fishing in a bay of which they knew nothing.

'With patience, we shall in time discover a great deal about this country. We already know far more about the people and their way of life than Tasman ever did.'

Could they not set up a camp ashore for the sick, asked Labé? It would enable them to get the full benefit of the 'land air' and rid them of the last salt in their lungs. With armed guards they need not fear any attempt by the natives to interfere with them.

De Surville refused but he did agree to change the anchorage, to bring the *St Jean Baptiste* closer inshore, reducing the distance the boats had to travel with the water casks. He was aware of the Maori's changing mood — Labé, who seldom went ashore, was not — and he realised that it would be wise to reduce the length of the stay in Doubtless Bay. Accordingly that evening the *St Jean Baptiste* struggled against the tide and wind to anchor in eighteen fathoms facing the little bay where the French obtained their water and which de Surville named Chevalier Cove — thus both the main backers of the venture would be commemorated, and the jovial governor of Chandernagore could console himself, if the profits were low, with the thought that his name would appear on the maps of New Zealand.

The wind freshened during the night. By early morning it rose to a strong south-westerly gale — driving the ship slowly towards the shore, for the holding qualities of the bottom were poor and the anchor was slowly dragging over the bed of broken shells and sand. All the available men were on duty. The wind began to scream through the rigging, and the ship, after the stillness of the past few days, was once again creaking and rolling. One squall after another struck her. A salty damp, mingling with the urine of the last few pigs, once more ran down the companionways and seeped into the 'tween-decks.

De Surville lowered his only large anchor, paid out more cable, and the ship responded. A grey dawn was rising over the land, heavy with clouds and the threat of chilling rain. No canoes came out to the ship and no Frenchman went ashore. The men rested from their exertions, watching the white waves clawing the base of the cliffs.

The next day was calmer, if still cool, but now at least the heavy clouds thinned out and parted to reveal long strips of blue sky. The sea had moderated a little and the bay once more was alive. Canoes slipped in and out of the coves; men went fishing, and on the heights, inside the forbidden *pa* and behind its high palisades, where the French had glimpsed strangely carved, red-painted huts and sinister looking idols with extended tongues and eyes of mother-of-pearl, the women went about their mysterious tasks and the old *tohunga* brooded.

The canoes turned and made for the ship. The chief himself was paying a visit, accompanied by a crowd of warriors and this time by a few women. The *wahines* puzzled the French. Few among them were well enough to feel any interest in women. No doubt the women found it difficult to understand why the white men should be so listless. The French wrote them off as lascivious and impudent. Their dances, performed that day on the forecastle, were hard to interpret. Their arm movements were graceful enough, but the shaking of thighs and stomachs less so. The men watched impassively or with mild curiosity, their indifference exciting the women and driving them to unequivocal gestures. Sir Joseph Banks, a few weeks earlier, had referred to Maori women as 'skittish'. Labé, had he known of this comment, would have dismissed it as a prudish British understatement. They were lewd; they were a nuisance; and they had to be kept away. He, like de Surville, was concerned about his men being led into some ambush. He had learned to distrust all 'savages' of either sex. Nothing very serious could happen on board ship, but if a man, once he had recovered his health and vigour, were to be lured into the bush by some native woman, murder might ensue; and if several Frenchmen were tempted into taking such risks a whole boatload might be massacred. The women, therefore, had to be driven back to their canoes with yells and threats and if necessary with blows.

'Dirty, ugly sluts!' he shouted — and for the benefit of the sailors: 'A man would have to be hard up to risk catching their lice.'

The men laughed and turned to watch the chief and his followers who had paid no attention to the scene. They were more interested in the guns. The metal, totally unknown in New Zealand, puzzled them as much as the possible use of such strange objects. Young de Surville attempted to explain their purpose. He showed them a cannon ball, brought it up to the mouth of a gun, said 'boom!' and raised the ball to his chest. Then he leaned back against the rail, his arms outstretched, his head on one side and his eyes closed.

The chief raised his *patu,* nodding his head and saying *mate* several times.

'Fire the gun,' ordered the Captain.

Two gunners loaded the gun, while de Surville pushed back the chief and, drawing a high circle with his arm, pointed to the open sea.

'He began to watch attentively, as well as those of his people who were with him on the quarter-deck. We fired and they were all very frightened. He was the one who showed the least fear, because he did not cease to watch attentively the spot where the ball was due to fall, and when he saw it causing the water to spout up to a great height he gave a loud exclamation and spoke to his men.'

Some of the canoes, however, were speeding for the shore. The sound of the shot had echoed through the bay, sending up flocks of frightened seabirds whose cries of alarm were now resounding along the cliffs. It had boomed through tall grass and between the pohutukawa trees; and in the *pa* above, the women had fallen silent and the voice of the *tohunga* was stilled. What terror lay hidden in the bowels of that strange ship, from which a sudden thunder had emerged? What dangers did the presence of these white men herald? Had there not been another similar ship — more heavily built, but equally alien — which had sailed across the wide mouth of the bay a mere fortnight ago? No such ships had ever been seen before — whatever the legends might say about some other great canoe being driven off by warriors many generations ago. But now, how many huge craft of this kind were there around the shores of Aotearoa and what upheavals and disasters might they not portend? The heads of the elders were bowed and a strange fear seized them, the echoing warning of the unknown still rolling across the great bay.

De Surville, having established the superiority of his arms, wished to cement his friendship with the chief. He ordered two of his most precious possessions brought to him: two pigs. The Maori had no meat to sell him, although they had offered him a dog, which he had refused, and it was clear to him that pigs would be of great value to a people whose main food was fish and root crops. 'I showed him our pigs and gave him a young male and a young female, trying to make him understand that if he kept them he would have many of them from their mating. This seemed to please him greatly and he took them away with every sign of satisfaction.' The sacrifice, for a ship whose food supplies were now tragically low, was a very real one.

By now the sea was moderate enough for the water party to return ashore. 'The chief came with me instead of going in his canoe. On the way he urged me not to land at the usual cove, but to make for another, telling me by signs that there is more fresh water there.' There was indeed a larger cove nearby with a deeper stream, but the beach was so pebbly that it would have been impossible to roll the full barrels over it. De Surville was also worried by the fact that the boats would be hidden from the ship by a headland and would not have the support of the guns in the case of trouble. He accordingly went back to his usual landing place — and there realised why the chief had tried to lead him away from it:

'We made for our first cove where there was a large quantity of freshly caught fish. I think that this was what they did not want me to see. Since they share it among themselves and live on it I was doing them harm.'

De Surville was again showing himself aware of something many other navigators did not understand — that the large-scale purchase of foodstuffs by a ship's crew can present a real threat to a native population living at bare subsistence level. On this occasion, however, the Maori agreed to sell a good part of their catch; nor were relations really strained because, when de Surville was again asked for his sword and, exasperated by this ceremony, refused to part with it, the chief was but slightly upset.

From almost every point of view the stay at Doubtless Bay had proved to be a success. Only a few more days were left in which trouble could arise, whether through Maori patience becoming exhausted or through some thoughtless action by the French — or by a set of circumstances no one could foresee or circumvent.

PART VIII
KIDNAPPED

On Christmas Eve, a Sunday, the weather was fine, with a light south-westerly. De Surville went ashore, once again accompanied by the chief. Everything was quiet. A little fish was bought and the sick, now improving rapidly, walked quietly under the trees. 'The local natives no longer crowd around us. They are beginning to believe that we mean them no harm, and are no longer on the alert.' The boats brought back their quota of barrels and firewood. A few canoes sailed around the ship, some with women — 'quite ugly', commented Labé unsentimentally — most of them with fish or greens, which were far more acceptable.

When night fell, Father de Villefeix in his Dominican habit read the evening office prayers and the customary prayers for the sick. One man down below had lost consciousness. Another, one of the unfortunate three captured in the Batan Islands, was not expected to live. Others, now able to sit up, were thinking of Christmases gone by, of the ceremonies and family dinners that were being prepared in Britanny and in Normandy, of traditional songs and special recipes. The yellow light of a candle glowed dully in a circle of gloom. Someone went over to the silent Batan Islander, he was still breathing. At 1 a.m. he was dead. Labé started a new page of his journal:

'Monday, 25 December 1769. Christmas Day. From 8 o'clock last night until midnight fair weather, some cloud, light breeze.' Then he drew his black cross.

The dead man was lowered overboard at dawn. Then the day's work began and de Surville went ashore with three boats. Was Villefeix with him or did the chaplain remain on board? More important, did he say mass?

It would be neat, clear-cut, if Labé or Pottier had made mention of the Christmas service in their journals, but neither they nor the Captain referred to it. Can one deduce from this that there was none? Even a reference to evening prayers on Christmas Eve is but an assumption, based on no direct evidence. Yet can one assume the opposite — that a chaplain, not listed at any stage among the sick, above all a Catholic priest, required to say mass daily, would have neglected to hold some kind of service on Christmas Day?

The absence of a definite reference in any of the journals is not evidence to the contrary, for routine matters of a non-navigational nature have no place in these logs. There are no mentions of captain's inspections, watches, meals, entertainments. Why should de Surville bother to state that Villefeix was doing his job, when he does not say it of Monneron or Dulucq? He would have been more justified in saying the opposite, if Dulucq or Villefeix had been too ill or too neglectful to carry out their ordinary duties.

Since it was not usual for a merchant ship to have a chaplain on board, except when missionaries or priests were being given a passage to a definite port, we can assume that Father Villefeix was on board the *St Jean Baptiste* because she was bound for a new mysterious land where a missionary was needed to bring Christianity to the natives. Father Villefeix was an essential part of the plan of exploration. De Surville's family had always befriended Catholic missionaries. One of these had caused the conversion, in 1699, of his aunt, Elisabeth de Surville, who, abandoning a life of frivolity, had founded the Community of the Good Saviour in Saint-Lô. In 1740, the Bishop of Halicarnassus, after staying with de Surville's own mother in Port-Louis, called her a true 'mother of missionaries'. Father Villefeix himself, in spite of one odd episode which if anything confirms the theory that he was a friend of the de Survilles — he deserted in Peru and went gold prospecting with the Captain's nephew — was a respectable priest. He had studied from 1760 to 1766 at the Convent of St James in Paris and had been appointed to a teaching post there. We lose trace of him in 1766, at which date de Surville was back in his home town of Port-Louis, supervising the fitting out of the *St Jean Baptiste.* Can we not assume that Villefeix, moved by a desire to do some missionary work, met de Surville when he was about to sail to India in mid-1766, and was later invited to join the great expedition? He was to end his life as a country priest, for we find a last trace of him in the parish register of Etouars, in Dordogne, where is recorded the death in February 1780 of 'Antoine Léonard de Villefeix, aged fifty-two, parish priest'.

Can we reasonably assume that Father Villefeix, who was known to be a priest in 1756 and to have died a priest in 1780, who is not recorded as being ill in December 1769, would have neglected to say mass on Christmas Day — merely because we have no direct statement to that effect from the ship's officers?

If we cannot, then Father Villefeix's mass is New Zealand's first Christian service, antedating by precisely forty-five years Samuel Marsden's famous service in the Bay of Islands on Christmas Day 1814.

It was probably said on board. A service on land would have been unusual and might have warranted that missing comment in one of the logs. But if the latter, then Villefeix would have gone ashore shortly after dawn in one of the two boats which took de Surville and the sick to Chevalier Cove; and in that case, Father Villefeix would have set up his altar on the sand or on the grassy slope, just above the water-line, no doubt in front of a small group of curious Maori. But it is more probable that he held his service on deck at first light, before the Captain left, for it is not likely that de Surville, a member of a devout family, and his predominantly Breton and therefore equally devout sailors, all of whom knew they might have to meet their Maker in the very near future, would have neglected to take advantage of a chaplain's presence on board on Christmas Day, and have used the excuse that there was a day's work ahead of them to avoid having to attend.

In the afternoon, de Surville, anxious to reduce his dependence on the Maori and the problems this was beginning to cause, decided to go fishing. Labé went with him: he had spent most of his time on board and deserved his breath of 'land air'.

'We took the seine with us because we were going mainly to see if we could find some convenient place to obtain fish without the assistance of the natives. ... We finally reached the cove at the end of the peninsula. We stopped there without rounding the point in order to throw the seine, the bottom being level and sandy. While M. Labé was laying out the catch, I, with a few of our gentlemen, climbed to the top of a large bluff which forms the head of the peninsula. It was there that I discovered it was not an island and that on the other side of this bluff there are some very attractive bays which would be very convenient for fishing.'

The catch, however, was poor. It improved a little when de Surville returned to the same bay the day after and again the day after that. But on the 27th, as he made his way back to the *St Jean Baptiste,* he found the weather worsening.

'As we left the cove we found the wind much stronger, and it freshened continually as we progressed. The yawl and the longboat in which I was were making headway, although with a great struggle; as for the other boat which held most of our sick, with the small Bashi boat in tow, it was dragging far behind. I promised a bottle of wine to my rowers if they arrived on board before the yawl which at that moment was a little ahead of us. This proved effective: we went ahead at once, in spite of which I did not reach the ship until 9 p.m. The yawl arrived soon after.'

The wind had grown into a fresh gale. A few hours more and it blew itself up into a wild storm. The two anchors started to drag. Labé lowered the main anchor with thirty fathoms of heavy cable. Rain developed,

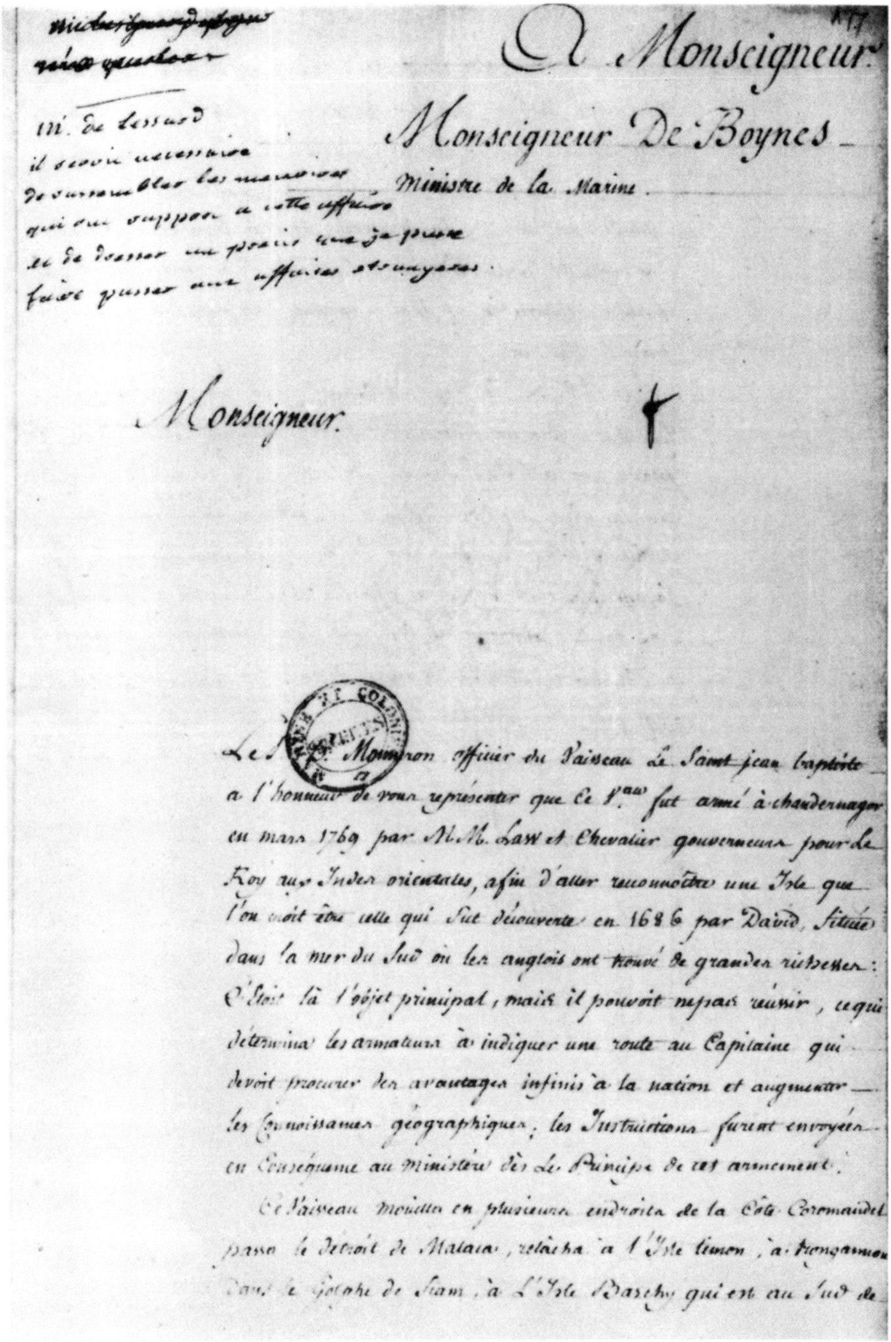
À Monseigneur

Monseigneur De Boynes

Ministre de la Marine.

M. de Lessard

il seroit necessaire

de rassembler les memoires

qui ont rapport a cette affaire

et de donner un precis que je puisse

faire passer aux officiers etrangers.

Monseigneur.

Le S.r Monneron officier du Vaisseau Le Saint Jean baptiste a l'honneur de vous représenter que ce V.au fut armé à chandernagor en mars 1769 par M.M. Law et Chevalier gouverneurs pour le Roy aux Indes orientales, afin d'aller reconnoître une Isle que l'on croit être celle qui fut découverte en 1686 par David, Située dans la mer du Sud où les anglois ont trouvé de grandes richesses: C'étoit là l'objet principal, mais il pouvoit ne pas réussir, ce qui détermina les armateurs à indiquer une route au Capitaine qui devoit procurer des avantages infinis à la nation et augmenter les connoissances géographiques; les Instructions furent envoyées en conséquence au ministère dès le principe de cet armement.

Ce Vaisseau mouilla en plusieurs endroits de la Côte Coromandel passa le detroit de Malaca, relâcha à l'Isle Timon, à Trongamon dans le golphe de Siam, à l'Isle Bachy qui est au Sud de

MONNERON'S REPORT ON THE VOYAGE

the gusts hurling it at the windows of the great cabin, stinging the faces of the men on watch. At dawn, de Surville turned his spyglass on to the shore, hoping to see a sign of his boatload of sick — for they had never reached the ship, but had been seen to turn back. Pottier was in charge of them. De Surville had confidence in him, but less in the Maori: would they attack the sick and their outnumbered, soaked guards? The land was shrouded in grey, the only signs of movement being the angry line of surf and the increasing rain.

At 8 a.m. the starboard cable snapped and the ship veered slightly, pulling at her two remaining anchors. A gun broke loose from its tackle and began sliding across the gun-deck. A loose gun on a rolling deck is like a terrified animal, dangerous and unpredictable. Labé went down with half a dozen men. They watched it like toreadors around a bull, trying to anticipate its sudden murderous lurches, then managed to throw ropes around it and secure it.

Heavy seas were sweeping the decks, pouring down the companion-ways, soaking the bedridden scurvy cases and running down to the hold where the water rose inexorably at the rate of an inch an hour —no one could be spared to work the pumps.

When Labé returned to the quarter-deck, de Surville pointed to a line of rocks which seemed suddenly to have moved towards them:

'We're less than a musket-shot away. Only one thing can save us. Raise anchor. We will sail out.'

When he gave his orders, the men seemed dazed.

'They lost their heads to some extent. Before carrying out my orders they stared as if they had gone a little deaf. Fear had seized them.' Like Labé they did not believe the ship could sail out of the bay in the teeth of a fierce north-easterly gale. But de Surville was an experienced seaman. The ship responded at last, straining fiercely, amid the crash of boxes and barrels which fell and rolled across the sloping decks, sailing so near the threatening rocks that Labé and his men saw death before their eyes, 'with the rocks passing alongside the ship close enough to make your hair stand on end'.

De Surville tacked about in the gale and slowly made for a bay a little farther to the west, which he named Refuge Cove, and lowered his remaining anchors. His first thought was to rest his men who had had no food since early morning. No one had taken any rest, no one had cooked any food, few had spared a thought for the time. They realised that dusk was beginning to fall, the light failing earlier because of the stormy sky. There was little more he could do. The sea was still high, with heavy waves breaking over the bow, but the wind was beginning to moderate. It lessened appreciably during the night, turning gradually more northerly so that the peninsula began to provide shelter from the worst gusts. In the morning, although the tiller was found to be damaged, it was clear that the *St Jean Baptiste* was out of danger.

Pottier de l'Horme brought back the sick as soon as the weather permitted. He had been compelled to turn back, reaching shelter after a long and hard struggle, his boat half swamped; he had gone up to the edge of the beach, but decided against landing. His men spent the rest of the night in the greatest discomfort, soaked and cramped:

'I required everyone to maintain complete silence so as not to be heard by the natives who live on top of the headland, for fear that they might think we were taking advantage of the dark to go and attack them by surprise. As soon as we were far enough inside the cove to be sheltered I lowered the grapnel and covered everybody, especially the sick, with the sail, then I got them to bail the water out of the boat and set up a good guard for fear the natives might appear.'

His report continues, careful and detailed:

'At first light I brought the boat up to the shore, got my people off and lit a great fire to warm them. Some natives appeared shortly after and came up to us. The chief of this cove was among them. I made him understand by signs that our return had been caused by the bad weather which prevented us returning to the ship, which he understood very clearly. As we had some fish with us which we had caught the day before, I shared half of it among all our people and kept the rest for the next day in case it was not possible to return to the ship, because this was all the food we had. But I was most pleasantly surprised when some kind of chief came towards me with another native who was carrying some dried fish. I made signs telling him I had nothing to give him in return; he said that he was not asking for anything and in addition offered me his hut to put us all up. This offer seemed to emanate from a truly humane heart touched by the accident which prevented us from returning to our ship; but I did not think it advisable to trust him.'

Unwilling to take any risks, Pottier was forced to spend another night on the beach, sheltering under a tent which he made out of oars, branches and the boat's sail. During the day, he and the men under his care had watched the drama of the *St Jean Baptiste's* struggle to escape destruction and rejoiced to see the ship finally coming to anchor under the lee of the western hills.

De Surville now had only two anchors left. He could take no more chances. In less than a fortnight spent off the New Zealand coast, two violent storms had assailed his ship, and he could not afford to wait for a third. Nearly all the water casks had now been refilled, he had a small stock of dried fish and greens, and although there were still a number of sick — indeed two died just after the storm and a third was to die soon after — the condition of the crew was generally much improved. He once again went ashore, taking Labé with him, intending to fish for the last time.

His journal includes an ominous addendum: 'and also to see whether I could not take away a local native to try and obtain from him later on what information I could about this country'. Now that he realised that he could not stay much longer in the bay, he had decided to repeat what he had done in the Solomons.

The Maori were again pleased to see them. 'The chief welcomed us in a most friendly manner. We climbed on a rise near one of the huts where they gathered eight or ten of them around us, including three young girls or women who danced in front of us for a long time, trying to attract us by all kinds of highly lascivious movements. Two young men also joined in their dance. Finally, bored with always seeing the same thing, because they would not desist, thinking apparently that we were simply difficult to arouse, but that they would succeed in the end, we left. … The three females followed us and finally, in a burst of excitement, one seized me around the waist and, holding me tightly in her arms, made some highly lascivious movements against me. I freed myself and we got on with our work.'

De Surville, with no understanding of traditional Maori dances and in no mood to appreciate the uninhibited enthusiasm of the girl, prepared to leave the cove. The wind had risen and he was afraid it might develop into a new gale and endanger the ship. Before he left, he gave the chief 'a handsome cock and hen, which were the only ones we had left'. To this gift he added peas and wheat, 'explaining how they should be sown'. It was once again a generous gift of items he could ill afford, and would have been a fitting end to his stay in New Zealand, for nothing had yet happened to mar the excellent relations established between French and Maori.

The trouble — what ruined his reputation and painted him as a blackguard — was his desire to take away a native from whom he could learn all he wanted about the country and its resources. The purpose of the voyage, after all, was to find new outlets for trade goods and some island suitable for the establishment of a French trading post. New Zealand might offer at least the latter possibility. If he had attempted to explain his wish to the Maori, they might have found him some youth eager to go with him, for the Polynesians were adventurous navigators, and both Bougainville and Cook had taken volunteers away with them from Tahiti. But the next morning —the last day of 1769 —an incident changed the entire situation.

The yawl, which had disappeared in the storm and was believed lost, was noticed stranded high and dry on the main beach west of the *St Jean Baptiste* and surrounded by natives. De Surville lowered a boat and hastened to reclaim it. The sea was running moderately high, so that neither the Captain, who was acting as coxswain, nor his rowers, were able to keep the yawl in sight. When they reached the beach, it was quite empty, with no sign of yawl or Maori.

The sands run for miles in a great curve to join the main sweep of Doubtless Bay. Low sand dunes back on to the beach, concealing rush-filled swamps and a meandering stream. De Surville, thinking he had landed too far from the place where he had first seen the yawl, walked first in one direction, then in another. Finally, he spotted tracks on the dry sand showing that the Maori had 'dragged it over the sand dunes which line the shore up to a little river situated on the other side of these same sand dunes. They had hidden it or sunk it, or perhaps taken it to nearby lakes to which this river seems to lead.'

The search had taken almost four hours. De Surville and his sailors were exhausted after walking back and forth among the sand dunes and swamps. They shot a few birds, collected some wild celery and made a soup there and then, 'a heavenly meal'. The appearance of Maori, keeping at a safe distance, but angering de Surville by their presence and their caution, so clearly due to guilt, reminded him of his wish to take a native away with him.

'I went towards the right-hand corner of the bay where the houses are situated. We had not gone a third of a league when we saw large numbers of natives moving about, some running up the hills, others running down, and groups of others sitting down on these same heights. Some of the older ones were not very far from us; one, coming forward, even gestured to me to come up, instead of which I stopped and myself gestured for him to come to me. He hesitated for a long time. In the end he made up his mind and came straight towards me, unarmed. I reproached him for the theft of the yawl and at the same time told some of the sailors I had brought for this purpose and who had ropes with them to take him prisoner.' After this, de Surville set fire to a canoe, some fishing nets, and a few fishermen's huts, and went back to the ship. Another canoe, hauled up on the beach, was towed to the *St Jean Baptiste* and hoisted on board to be taken to France.

The kidnapping of a single Maori would have been relatively unimportant — there was no ill-treatment — if he had not been the very chief who had offered food and shelter to Pottier de l'Horme and his boatload of sick, marooned by the storm. Pottier recognised him at once. 'I was touched by the greatest compassion,' he wrote, 'when this unhappy wretch came on board who, recognising me and not knowing what his fate was to be, threw himself at my feet, clasped his arms around my knees and then rose to embrace me with tears in his eyes, saying things I could not understand, but telling me by signs that he was the man who had brought me the dried fish when neither I nor those who had been unfortunate enough to fail to rejoin the ship had anything to live on. This man seemed to be pleading for his life, or for me to plead on his behalf. I did what I could to console him and make him understand we meant him no harm. It was useless, for he did not stop weeping, especially when he saw his feet being shackled to prevent his escape.'

A victim of ingratitude always evokes sympathy. Dulucq, who had been with Pottier, also urged de Surville to release him, but de Surville was not prepared to waste time on the probably futile task of trying to catch another Maori in his place. Labé was less sentimental:

'The Captain would have done better to capture a dozen of these islanders to help us work the ship. We have lost sixty men, and forty others are in poor health.'

Dulucq persisted that they should release him and go ashore in search of some other native, but he was firmly overruled. Such sentiment might lead to a clash, said the Captain, perhaps to the death of a French sailor and as an inevitable consequence to the death of several Maori.

The men were already at the capstan; the anchor was slowly rising from the sea bottom. Others were up in the rigging, awaiting the whistled and shouted orders from below. A fresh south-westerly was blowing, with light intermittent rain. The *St Jean Baptiste* turned and tacked to stem northeast. She rolled and pitched as she left the shelter of the bay, then settled on her course with a following wind. It was 10.30 p.m. on New Year's Eve 1769.

PART IX
THE PACIFIC

As the *St Jean Baptiste* prepared to sail out of Doubtless Bay, its shores hidden in a darkness made more impenetrable by the drizzle, all the officers were assembled in the ward-room, arranged according to rank and character. The Captain was at his desk with a rolled-up chart at one end and his boxes of papers at his elbow. Labé was sitting, awkward and stiff because of his wounds, half facing him on his right. Monneron was seated at a small table, a blank sheet of paper before him, his quill pen at the ready, serious, self-effaced, efficient. Pottier stood a little to one side, knowing his place, keeping his own counsel, with Dulucq leaning against a wall cupboard, impassive; Villefeix near him, withdrawn, his eyes downcast, his breath smelling just a little of brandy. And at the back, away from the oily glare of the lamp, the younger lieutenants, Hughes de Surville among them, quiet, a little awed, and the boatswain, stolid and determined, never speaking unless spoken to.

The last act was about to begin. Fate is nothing if not ironical, and irony is best achieved if men determine their own destruction, making use of that free will in which they put their trust.

De Surville spoke first:

'We must decide, gentlemen, what steps we are now to take. Our crew and our ship are not, to my mind, in a fit condition to continue a voyage of exploration without first calling at some port where we can refit and more satisfactorily refresh our crew. We have, as I see it, three courses open to us: to make for the Philippines or at least for Guam, to sail to the Dutch East Indies, or to carry on to South America. We might still find some unknown island where our people could rest, but I doubt whether we could find any equipment for the ship.'

This, said Labé, was the real question. Where could they find the stores they needed — or the men to make up the crew? Not among savages. They had lost a third of the crew — he had drawn sixty-two black crosses in his log so far. Sixty-two savages, nor yet a hundred, from some wild island would not make up for the loss of trained sailors, of carpenters, caulkers or sailmakers. Nor would they find new sails, new rigging, the cable and the anchors they so badly needed, unless they reached a civilized country.

De Surville turned to Pottier, who confirmed that neither the ship nor the men could stand a long voyage. They could hold out for two months more, perhaps three, but not four.

The alternatives were discussed one by one. Anything north of the Line — Manila or Guam — was ruled out. They could not return to the tropics: the heat, especially if they were becalmed once more, as was likely, would be fatal to the men.

It was no easier to make for Batavia, Amboina or Timor. They would have to beat their way up along the coast of New Guinea, in seas rumoured to be filled with coral reefs and unmarked islands. Their charts were wholly unreliable. The existence of Torres Strait, which offered an alternative route, was uncertain and in their condition it would be folly to take a chance.

They were left with South America. It was relatively near, and time as they all knew was what mattered most. The winds in their present latitude were usually westerly and therefore favourable. With luck they could reach Callao within two months, on a steady course, with little manoeuvring.

Besides, they were not beaten yet. Most of the Pacific between New

Zealand and South America was unknown. They might yet discover some exciting new land, at least one where they could refresh their men and get food and water. It was true, as Labé pointed out, that so far their ports of call had been a great disappointment, but there could well be a better country lying somewhere along the thirtieth parallel.

De Surville issued a warning. The Spanish had always forbidden foreign ships to call at any of their ports along the Pacific coast. They upheld a policy of exclusiveness in all their American colonies and anyone who ventured near was liable to be imprisoned. The instructions issued to him enjoined him not to call anywhere in Peru or Mexico.

Labé was silent. The instructions! He for one had never seen them. He had risked his life and his money, and might well be about to lose both, without having been told what the instructions contained.

But what was the alternative?

'It is the only course left to us,' he wrote in his journal later that day, 'although a forbidden one, but being allied to the Spanish crown and knowing that nation for its feelings of humanity, we have great hopes from the gentlemen who rule there, and in that way we shall be able to restore the health of our crew, repair our vessel, and obtain supplies of cable and anchor, food, sailors, &c, in order to leave promptly to continue our search for new discoveries, and to return to Pondicherry.'

Monneron drew up a memorandum summarising the discussion, which they all signed. A course was set for northeast, and the *St Jean Baptiste* sailed into the unknown, with nearly seven thousand miles to go before her next port of call; her seams, ill-caulked, wept continuously, she creaked and rolled, her hold filled with goods which so far had been of no value and which were now mildewing and slowly rotting in the continual dampness; an ark, heavy with misery, struggling through unknown seas towards a goal which would bring relief only if the jealous laws of Spain could be stretched to include pity.

The daily routine of life at sea returned. It was to last over three months, with no sign of land until the end of March. Pottier had reckoned that the ship and the crew would not hold out for more than four months. His theory was not to be put to the full test — by a mere three weeks. But by then it would be clear that his estimate had been correct.

The men at least were better. There were a number of sick below, but not a tragic number. 'Land air' and New Zealand cress and celery had worked wonders. The sailors were still slow in their movements, hesitant, easily depressed, quiet, as one could expect men to be who had stared death in the face and who lived in dread of its return. The ship, however, could still be worked without inflicting hardship on those who were fit, for the winds were steady and the course straightforward.

The Maori captive, whose name was Ranginui, soon got over his sorrow and his fear. De Surville had a bed set up for him in a corner of the ward-room which he shared with the young Solomon Islander. Both dined with the Captain. Of Ranginui, Labé could write on 1 January: 'Our islander from Lauriston Bay seems to have resigned himself to the inevitable. He no longer looks sad, he laughs with everyone, eats and drinks well and sleeps well.' Young Lova, the youth from Port Praslin, had by now become accustomed to shipboard life. He spoke French passably and kept the sailors' spirits up with 'his pranks and monkey tricks'.

Labé opened up some of the bales and hung the cloth to dry in the warm sun. Water damage was not as bad as he had feared, but water continued to seep through into the hold. The cargo was restacked and part of the daily routine was an hour at the pumps.

At midday the officers made their daily observations and worked out their estimated position. The latitude was fairly correct — lessening gradually as they continued on their north-easterly course; the longitude was largely a matter of guesswork. It was of little importance since they were sailing across a total blank on the map of the Pacific — if they sighted any land they would be its first discoverers. They kept watch therefore for any indications of land — birds, seaweed, floating branches. Six days away from New Zealand, Pottier saw a flight of butterflies. Where did they come from? Where indeed! But the horizon remained empty and the

RANGINUI

telltale swell rolled on day after day from the south-west, making it plain to all but the most optimistic that there was no land to starboard.

The afternoon was a long dreary interval between meagre lunch and meagre dinner. The dried fish ran out, the last of the greenstuff was eked out; both were used for soup, the standard dish and the cook's great stand-by since no one could be sure what it contained and it enabled nothing, however putrid, to be wasted. The day would soon come when he would hesitate whether to skin a rat and add it to his stew. Fortunately in a way, for it broke the monotony, Labé could set his men to repairing the sails. They were splitting almost daily and there was no alternative but to patch them, for the spare sets had long since been used up — the *St Jean Baptiste* had been at sea for over eight months.

Early afternoon was also a time when most journals were written up. De Surville's entries were brief. Like the others he had, shortly after sailing, written his impressions of New Zealand and its people — although nothing as detailed nor as valuable as Pottier's long careful 'observations'. But after this he wrote a brief paragraph only, and one notices unusual little errors: the strain was beginning to tell and his physical condition was becoming impaired. Labé's journal was once again his confidant. Discipline stopped him from sharing his anxiety or his complaints with either Pottier or the Captain. Food, of course, was important:

'Our islander from Lauriston Bay and our little Kaffir captured at Port Praslin … are the first to be served at table and get the best, which means that there is nothing left for us.' Another day he clings to hope: 'God is great. I hope He will not abandon us! His will be done.' Then he records his fears about the state of the crew: 'Our sailors, convalescing from scurvy, are starting to complain of pains in the chest. I am afraid it will not be long before they fall ill worse than they were before we arrived in New Zealand. God's will be done.'

When he wrote this it was still only 9 January. There are no special comments for a few days, then he expresses concern at a shifting of the wind: the *St Jean Baptiste* was compelled to change to a south-easterly course, the first of a series of great zigzags across the South Pacific. Worse still, the unfavourable winds reduced the ship's progress to almost nothing — five miles on 17 January. By the end of the month Labé's health began to worry him:

'We are weak because of insufficient food, and I am beginning to feel so weak I find it tiring to stand up, without being actually ill.' But he, unlike the ordinary seamen, still had a small private stock of food. The condition of the ship was causing him increasing worry — she was now making water at the rate of five inches an hour. Would this exhausted, leaking ship ever reach South America?

Although he continued to bring the bales up on deck to be opened and dried out, the work was as futile as it was exhausting, since the bales were soaked again by the next morning. On 11 February the water rose eight inches in an hour. 'The water pours down into the hold from all sides.' Things improved a little when the sea moderated, but on the 14th, he wrote:

'We have been suffering from hunger for a long time. No lunch, a thin soup for dinner, and no supper.' Worse still, some of the water barrels had leaked, probably through being rolled too roughly on the beach, and eight were found to be empty. The water ration was accordingly reduced.

And now, inexorably, inevitably, the number of sick grew. Labé resumed his daily visits to them. 'I am continuing to supply our poor sick with some items out of my little store, sugar, tea, coffee, jam, syrups, wine.' On the 20th, he once again drew his thick Maltese cross. 'At 4 p.m. yesterday one of the islanders we captured at Bashi died of scurvy. He was about twenty-five. Only one is left out of the three captured at Bashi.' The next day the only remaining caulker reported sick; there was no one left to cope with the ship's gaping seams.

Pottier made fewer comments during February. He did allow himself one marginal note on the food situation:

'We have now only one meal a day and that consists of dried fish served up in a stew.' But at the end of the month both he and Labé came to the conclusion that de Surville had been hoping to land in David's Land, and that possibly this was in fact the main aim of the voyage.

'It seems from the course set by the captain last night that he is attempting to reach the latitude of the land or islands seen by Captain David, an Englishman, in 1686, and then follow along its parallel to try and sight it,' wrote Labé, a view confirmed by Pottier in his entry of 28 February, two days later.

But by then de Surville had, for the second time, taken Labé into his confidence. He was indeed hoping to find David's Land. There had been no sighting of a single island during the previous two months, and it could be presumed that these uncharted seas were empty. But now they had reached a region which had been visited before and where land had once been reported, more than eighty years before, by one Edward Davis — not David — who had told his fellow-buccaneer William Dampier of his discovery. No one had ever seen this island again and most geographers now agreed that it probably did not exist — as Labé now told de Surville:

'Personally, I am convinced it was only a cloudbank seen from a distance. Several navigators have looked for it, but none has ever seen it. It is either a cloud or else it was that Easter Island discovered by the Dutch.'

'Our information, M. Labé, is that this island is precisely the one rediscovered by the English last year.'

Was this then really the place they had been sent to find?

It was, said de Surville. There had been confusion over names. Their informant in Cape Town had mentioned an island discovered by an Englishman called Vallis, but the correct name was Davis. The island in question was often wrongly called David's Land; it should be Davis Land.

A confused Labé reminded him that an English expedition had set out for the Pacific two or three years previously with two ships and two captains, one of whom was certainly a Captain Vallis or Wallis. De Surville swept all objections aside.

'I can assure you, M. Labé, our informant is a reliable man. Otherwise,

I would not have risked my fortune on such a voyage, nor would my partners have risked theirs.'

Labé could do nothing but look thoughtfully at the white clouds scudding overhead. He had to admit that de Surville knew a great deal more than he did, for indeed he knew very little. But Davis Land? He found it hard to swallow.

And yet it was correct, in a way; it was out of such a hotchpotch of rumours, mixing up Davis's rumoured sighting of 1686 with Wallis's discovery of Tahiti in 1767 that Law and his friends had created the mirage towards which the *St Jean Baptiste* was still painfully sailing.

'Our supplies won't allow us to tack about in these waters for very long.'

'Unhappily. I will spend another few days on this search, although of course each day will still bring us a little nearer South America; if we discover nothing we shall sail to Peru, refit, and then return here. Because I am convinced this is where we shall find it.'

There were now thirty-eight barrels of water left. Although nobody could know, another five weeks would go by before they reached Peru. The water ration was cut back once more. The hope was that it would last a month. Labé wrote:

'I am so exhausted I can stand only with difficulty, because of lack of food. The crew are in desperate straits and are beginning to suffer a relapse, each being affected to some extent by scurvy. God knows what will become of us.' The next day one of the water casks was tapped for the daily ration — it was empty. In the evening the winds shifted and a light rain fell which grew overnight into a steady downpour. This was a blessing because it enabled containers to be placed on deck to catch the rainwater dripping from an awning, but visibility was down to half a mile: if the *St Jean Baptiste* had been sailing along a coastline, no one would have seen it.

Finally, de Surville again called his officers together. He told them what they had all guessed, that he was hoping to find Davis Land. The meeting was short. He recorded the outcome of it in a brief note in his journal:

'We could no longer amuse ourselves looking for Davis Land but had to go as speedily as we could to the Spanish coast on account of the shortage of water and firewood and the sick.' It was Wednesday, 7 March 1770. The martyrdom would not be over for a month.

A lascar from Bengal died during the night. The burial ceremony was short. The crew were too exhausted to pay their last respects to the dead man. Labé was having bouts of fever and his back was increasingly troublesome. 'I find it painful to sit or to rise.'

But two bonitos were caught the next day. 'A great feast for today and tomorrow. May God grant us such a favour every day!' The feast was not repeated, however, for many days; shoals of small fish did appear, but none was caught. Meanwhile, 'the entire crew is sick'. On 12 March the horizon was at last broken. The French saw a ship for the first time in seven months. It was small and after a while disappeared to the south, but it was at least a herald of civilization.

'The sick suffer terribly,' wrote Pottier on the 18th. 'They are rotten with scurvy,' added Labé. According to his reckoning he should have been almost within sight of Juan Fernandez Islands off the coast of Chile, but of what use could dead reckoning be after such a long voyage? Ranginui was now very ill — 'I consider him a doomed man. He refuses food, and scurvy affects all his joints'; but he was not the next to die. One of the soldiers, a native of Toulouse, died on the 21st; the popular 'Sails' followed him the next day. 'Our men are continually collapsing from exhaustion. I brought up 230 bottles of wine from the main hold to give them two rations daily. I also brought up a cord and a half of firewood. It will last as long as it can. I do not know which saint to turn to.' But on 24 March, the look-out called, 'Land ahead!' The *St Jean Baptiste* had at last reached the islands of Juan Fernandez, lonely outposts of South America. For some they no longer had any meaning: Ranginui died at nine o'clock that same night.

De Surville tried to send two boats ashore for water and firewood, but the sea was running too high. And even if some of the men could be coaxed to man the boats, who would remain to work the ship should a storm blow up? 'There are only five men left on the forecastle and these have scurvy.' The *St Jean Baptiste* sailed on. Help would be needed from the Spaniards not merely to refit the ship and replenish the stores, but even to bring the ship into port.

'Thursday, 29th March 1770. Death of Juan Lopez, cook. Friday 30th. Death of a lascar from Bengal. We have three men on the quarter-deck and four on the forecastle, and even these have scurvy. Sunday, 1 April. Death of a servant boy from Bengal, aged twelve or thirteen.' Labé drew his black crosses almost daily.

'The state of our wretched sick is painful to behold,' wrote Pottier, 'and caused by a little thing like having no meat to make them a broth, but their number is so great they have no cot nor even a mattress to lie on. They are stretched out on heaps of tow. We have not the slightest thing to give them. It is hot and they are thirsty and have only a half-litre of water to make an infusion. They continually call out that they are thirsty.' And he might have added comments on the stench and the darkness, the constant tossing and pitching of the ship, painful swollen limbs, and the all-embracing sense of despair. What could Dulucq do, his store of remedies now exhausted, except try to console them and pray for a favourable breeze? We lack Dulucq's journal, so we cannot tell what his anguished comments were, but another ship's surgeon, Taillefer, wrote in similar circumstances thirty or so years later: 'Nothing was more hideous than the appearance of their face: to the leaden complexion of the victims of scurvy was added the prominence of the gums jutting out of the mouth… The sick gave out a fetid smell, which, when you breathed it, seemed to attack the very root of life. I have often felt my strength ebb away when I came near them. Their state of weakness did not prevent them from retaining the use of all their intellectual faculties, which made them feel all the more cruelly the pangs of despair.'

De Surville, alone by the rail, watched tensely as the coast of Peru hove into sight. Another two days would bring him to the port of Callao, where he would find food and water and hospitals, where men lived a life so far removed from the nightmare of his ship that they seemed like inhabitants of a remote, almost imaginary world. But how many men would die in those two days? Dulucq had warned him that eight of them could hardly be expected to live. What could he do? The coast was empty and barren. An occasional fisherman's hut, a rare wisp of smoke, faint signs of human habitation, held out no hope of adequate supplies for the ship's starving crew.

As dusk began to fall on 7 April, the French came within sight of a village, with a whitewashed church set against the dark of the rising hills. It was not yet Callao, but Chilca, the nearest settlement south of Peru's main port. De Surville anchored off the bay and at first light fired a gun to summon help. There was no response.

He called Labé, handed him a letter addressed to the Viceroy, begging for help in the name of humanity, and asked him to take it ashore. Labé had the small, manoeuvrable Bashi boat lowered and set off for the shore.

An hour later, he was back.

'There's no hope of landing,' he said. 'The sea is breaking on a bar right across the bay. Our boat would capsize for certain.'

Dulucq was on deck, motionless, a symbol of his own impotence, staring at the village, with its neat little houses, its church, the groups of people one could just discern staring back at the ship. Below, the sick were groaning, calling out for help in one awful confused murmur. One of them would die during the night: Pierre Leroy, a Breton, aged twenty.

'I will go myself,' said de Surville.

'It's madness, Captain!' Labé exclaimed. 'The bar is far too dangerous.'

De Surville looked at him. For months he had carried the responsibility on his shoulders, a burden which had aged him and bowed his once proud shoulders. He had watched his men die one by one, and every day the empty horizon, the leaking ship, the putrid food, the cries of the sick, had crushed him yet further. Many reproaches would be levelled at him, but never cowardice. Labé lowered his eyes.

He dressed in his blue uniform with the gold braid and the white cross-belt. He pinned on his chest the Cross of St Louis and buckled on his sword. Tomorrow he would have to face the Viceroy of Peru, not as a simple merchant who had infringed the laws, but as a French captain and gentleman.

Labé watched the small craft row away. It neared the edge of the bar, a frail basket of hopes bobbing on the huge waves, and then the great rollers hid it from his sight.

He never saw de Surville again.

The men of Chilca, who had been watching the strange vessel from the shore, later went down to the beach and gathered the body, with its soaked uniform, the gold braid torn and soiled, the Cross still pinned to the jacket, the sword still in its scabbard; the village priest recited the customary prayers, and the body was lowered into a grave, which perhaps some pious village women visited for a time, and which was then forgotten and lost forever among the tangled weeds and the rank grass.

De Surville had indeed reached the bar and finding it to be, as Labé had reported, quite impassable, he had placed the letter to the Viceroy inside a bottle, which he had sealed and tied around the neck of a lascar seaman. The man, with great difficulty, managed to swim ashore. And while the Captain watched his frightened messenger struggling across the bar, a roller had come up from behind and capsized the little boat. Neither de Surville, in his full dress uniform, nor the two rowers who were with him had been able to keep afloat for more than a few moments in the angry waters.

Thus it ended, almost expectedly, for de Surville. The dreams of profit and discovery had dissolved in the agony of disaster. Foolish they were, those dreams, based on gossip and the misinterpretation of rumours; and costly for the wretched sailors whom ignorance and poverty had led thus far; but so much had been redeemed by the sacrifice of a captain who — however doomed to failure his final attempt may have been — was prepared to risk his life to get help for his men.

PART X
END OF AN ODYSSEY

10 April 1770. The *St Jean Baptiste,* her paint long since faded and peeling, her sails patched with great, ill-sewn, irregular squares, was manoeuvring slowly and weakly off Callao harbour. The men, nearing the end of their long nightmare, were listless, unresponsive to orders. Whatever remnant of hope and willpower had driven them across the Pacific now vanished as the reaction set in. Labé, still aghast at de Surville's death, walked slowly back and forth along the quarter-deck, watching the men, reluctant to take over full command and drive them to a last effort.

He had never visualised the ship making port by herself. To reach the harbour-mouth seemed as much as she could ever hope to do. Someone else, some other force, would have to take over from there. And indeed, as he watched, three longboats filled with Spanish sailors rowed out towards the French and took over the vessel.

Several other boats followed, loaded with food sent by order of the Viceroy. It was a good omen. But in one of the boats there also came a detachment of soldiers 'under a certain M. Demetrius Egan, a lieutenant-colonel, sent from Lima to make sure no goods were unloaded or sold'. Egan was to become Labé's *bête noire,* 'a monster' in fact who was to plague him for many months and destroy his hopes of saving the expedition from final collapse.

For the present, Egan had certain duties to carry out which Labé, knowing that all Spanish ports along the Pacific coast were banned to foreign ships, could not oppose. He agreed to call the officers and crew together and warn them that anyone caught attempting to sell any goods or personal belongings, or even to land anything without specific permission, was liable to ten years in prison; but he was perturbed to discover that the French would also be required to pay Egan the sum of twelve piastres a day for as long as the ship remained under guard. Had he known how long this was to be he would have been horrified.

For a while Labé turned to the urgent problem of the sick. Forty-four of them were taken ashore to the hospital at Bellavista, a former Jesuit college. They were well treated, the wards were clean and airy. The Spanish, so far, were proving to be as humane as anyone could have hoped. But for some, it came much too late. Three men died the first night and eight more were to die at various intervals during the following fortnight.

When the command had so suddenly devolved on him, Labé's ignorance of the syndicate's real intentions had placed him at a great disadvantage. His petty complaints about the ship, the stores, and the food, paled into insignificance beside the problems now facing him. His sorrow over the Captain's death was genuine enough — 'I have lost a true friend,' he wrote — but it had thrust a worrisome inheritance on him. Egan left him no time for mourning, no time for collecting his thoughts. The Spaniard followed him everywhere, placing seals on every cupboard, asking him questions about the expedition and its intentions. Labé could scarcely admit that he was not sure, because this would at once have aroused suspicion; nor could he look through the Captain's papers for fear of finding some incriminating document for Egan to pounce on.

It was at this stage that Monneron came into his own. Monneron, as Labé had always suspected, knew a great deal. He had been in the Captain's confidence, he had links with the syndicate in Pondicherry, he was a man of business. The morning after that first sleepless night in Callao, Monneron came to his cabin with a notebook given to him by de Surville for safekeeping. Labé opened it and found to his horror that it contained a detailed plan for a trading voyage to Acapulco and Peru, in direct contravention of Spanish prohibitions. This, coupled with the *St Jean Baptiste's* arrival in Callao, could easily be seized upon by the authorities as evidence of intended contraband, and lead to the impounding of the vessel and the arrest of the entire crew.

They tore the book into shreds and cautiously crept into the galley to burn it: Egan had to be prevented from even suspecting they had destroyed some documents. They then hurried to de Surville's cabin and feverishly searched through his papers. They found another proposal for trade along the Mexican coast, and once more they crept down to the galley and burnt the evidence.

Egan arrived shortly after and pompously presented them with an order signed by the Viceroy, requiring all the ship's papers to be handed over to him for examination. The ship was then surveyed and a report drawn up on her condition and the repairs that would be necessary before she was fit to sail from Peru. A few days later the sails and the powder were taken ashore and stored; then after a further delay the cargo was unloaded. By the end of July the *St Jean Baptiste* was an empty shell. Demetrius Egan had gone at last — 'there was nothing left on board for the rats,' Labé wrote viciously. He himself had not been inactive: he had paid several calls on the Viceroy and assailed him with written protests. Never a patient man under the best of circumstances, he fumed against the Spaniards who prevaricated day after day, insulted him and robbed him.

The Viceroy had dismissed his complaints by stating that it was believed the French had a considerable quantity of pearls and diamonds hidden on board. Labé pointed out that none of the ship's records made any mention of such a cargo. 'He replied that I had no doubt hidden or destroyed the said accounts. I did everything I could to dissuade him, but he refused to believe anything I told him.' A thorough search was made but nothing was found. All through June and July Demetrius Egan could be seen at any time of the day looking behind cupboards and walls for likely hiding places, or questioning the crew. 'The man is truly a monster,' reported Labé. The officers could not go ashore without being searched. The chaplain waxed indignant because a small writing desk he wanted to bring away with him had to be almost taken to pieces before Egan would let it go. Labé, about to step into a boat, was stopped for the customary search; he replied that he would return to the ward-room, and that if Egan attempted to search him in front of his sailors he would run his sword through 'the infamous wretch's body'.

One day Monneron ran up to him out of breath. He had just learnt that the local prosecutor was about to lay a charge against the French for breaking the monopoly laws, and was blandly asking for the death penalty for all concerned, in addition to the confiscation of the ship and cargo. They composed together yet another letter for the Viceroy, but it is unlikely that he read it — for Labé was still sending him two or three appeals a week. It is probable that neither the prosecutor nor the Viceroy was particularly serious about the charge or really concerned about the *St Jean Baptiste.* They had — as Labé was beginning to realise — referred the matter to Madrid. This was far safer and far easier than making a decision themselves. The delay would enable them to squeeze all they could out of the French who, since they were forbidden to sell any of the cargo, were forced to borrow money for their daily needs. The legal fees were also mounting. Labé had engaged the services of a local lawyer who was more enthusiastic than wise for, urged on by the Frenchman, 'he expressed with too much freedom his opinion on the laws of navigation' and found himself in trouble with the authorities. Labé was forced to defend his own lawyer — a cause he took up with his usual indignant vigour, writing to France to enlist the government's support.

After six months of this, Labé decided to send Monneron to Europe. There was no alternative: in Callao he was at the mercy of the Spaniards, who dismissed his complaints with an irritating courtesy or an even more irritating curtness. What was the use of fuming or fulminating any longer? Somewhere in Madrid some clerk held the *St Jean Baptiste* file, yawned over it, and set it aside for another day or another week. Monneron, now the business manager of the enterprise, if such a term can be used, had contacts in France and through the Law and Chevalier families could put diplomatic pressure on Spain.

'I have given M. Monneron a copy of all the legal proceedings to be handed to the French Ambassador, with some curios from New Guinea and New Zealand, and the young negro Lova from Port Praslin, with instructions about what is to be done, and numerous letters for our ministers, directors and agents, with passports for him and the servants.'

This done, Labé resigned himself to a long wait. One day, he watched the departure of a Spanish expedition leaving in search of Davis Land. The Spaniards had at last accepted his claim that he had sailed on a voyage of discovery; so now, while the *St Jean Baptiste* lay at anchor in Callao, they sent two ships on a similar voyage, to forestall both the British and the French. When they returned in March 1771, Labé drew up a report and set down his private — and perfectly correct — belief that they had merely rediscovered Easter Island.

A year went by. Life had settled down to a strange empty routine. Labé had little to do, except to rage against the Spaniards, write letters of complaint, and enter deaths and desertions in his log. Morale, not surprisingly, was low. 'I get no news from France. It would seem that I have been abandoned.' The lure of the gold mines or mere boredom caused most of the desertions. The assistant-surgeon disappeared in July 1770. The chaplain fled with young de Surville in September 1771; recaptured a month later, they escaped again in January 1772.

In August 1772, twenty-eight months after their arrival in Callao, the French finally received some news from France. 'I received two parcels of letters from the clerk, M. Monneron, one from Madrid, the other from Cadiz. This is the only mail I have had from Europe since I arrived in the port of Callao. He informs me by these letters that the ship and the cargo are to be returned to us.' Monneron himself arrived at the end of the month, bearing the official release documents. But there were no letters for Labé, neither from the members of the syndicate, nor even from de Surville's widow 'after I had written more than twenty letters to each of them. I cannot understand why they should have forgotten me to this extent.'

There were further mortifications in store for him: the instructions were addressed to him and to Monneron jointly. 'I am mentioned in each clause, which reads: M. Labé and M. Monneron will do such and such a thing.' No doubt other letters than his own had been sent from Callao to France — young de Surville may have criticised him when writing to his family — but it took more than slights to weaken Labé spirits. He was not going to be ordered about by Monneron — in fairness to the clerk, there is no evidence that he ever tried to — and he was certainly not going to share his command with any landlubber. 'This will not prevent me from continuing to act as I have done in the past for the good of the expedition and to remain alone in control of all operations.'

Now, after the enormous gulf of inaction into which he had been forced for so long, life returned to normal, and he sprang back into activity. There were calls to be paid on the Viceroy, more letters to write, orders to be given for repairing the ship (for, in spite of the survey two years before, nothing had been done), the cargo and sails had to be brought out of storage. During all of September and October, carpenters and caulkers swarmed over the ship; the sound of saws and hammers, the shouts of boatmen and slaves, of officers and supervisors, brought unaccustomed life to the French merchantman which had lain at anchor for so long, almost forgotten, in Callao harbour. The Spaniards came down to watch, lining the quay, sometimes calling out instructions or encouragement to the workers.

The cargo was loaded and stored under supervision, for it was as important as ever to prevent trading or smuggling; but this time, fortunately, the officer in charge was not the 'infamous' Demetrius

Egan. Labé nevertheless had to agree to a detachment of Spanish guards staying on board until the ship was ready to sail — their pay being met, as previously, by the French. Finally, on 7 April 1773, three years to the day since the *St Jean Baptiste,* her crew on the point of death, had limped into Callao harbour, the expedition was ready to sail for France. 'At last, thank God, I am free of the Spaniards!' exclaimed Labé. He was almost free: after all the deaths and desertions there were not enough sailors to man the ship, and he was forced to take on an additional crew consisting of sixty-four Spaniards, all of whom insisted on being paid in advance. If a little lazy by nature and inclination, they were, he admitted, good sailors. They understood their work and they understood discipline. When they complained of the food, a fortnight out of Callao, 'I fell on them with big blows of my stick'. Food! What did these men know of food who had not sailed across the Pacific, who knew nothing of the daily ration of uneatable stew, of crumbling biscuit out of which maggots dropped wriggling on to one's plate, of meat that had shrunk to a thin salty leather which no amount of soaking and boiling could soften, of muddy water, eked out like a rare wine, which stank of all the fermenting impurities gathered in the creeks of distant and savage countries? What had they to complain of, these Spaniards, paid in advance to work in a ship that was certain to spell financial ruin to all those who had risked or lost their lives in her? And did not these complainers belong to a nation which had held the *St Jean Baptiste* prisoner these three years, to complete the ruin of all the investors, to squeeze every last ounce of profit out of French flesh and French suffering, to prevaricate and rob and cheat?'

The blows fell across Spanish shoulders. 'Food? Rogues! Carrion! I'll give you food that's fit for you!' And the blows made up a little for all his frustrations, for treachery in the Solomon Islands, for the horizon that had been empty on so many days, for the powerlessness of man in the midst of unknown seas, for Demetrius Egan, for the bored Viceroy, for officers who dared to desert.

There were no more complaints. The troublemakers became 'as mild as sheep'.

The *St Jean Baptiste* made her slow way down the coast, rounded Cape Horn far to the south, out of sight of land, and sailed into the Atlantic. The months went by. On 20 August 1773 she reached Port-Louis in Brittany, and dropped anchor in the spacious harbour beside a dozen merchantmen whose adventures had been neither so glorious nor so tragic. The long circumnavigation was completed.

Of the original number — one hundred and seventy-three — who had sailed from India, sixty-six reached France. Twenty-eight had deserted. Seventy-nine had died.

A group of silent women stood waiting on the quayside, under the chestnut trees. Their black dresses, the black shawls over their heads proclaimed them to be widows. One stood a little aside from the rest, also in black, but her clothes of a richer material. She was de Surville's widow. Unlike the others, she still had a few jewels to sell, some furniture, a little property, to save her from destitution. After a series of appeals she would be, one day, granted a small state pension and allowed to sink gently into humble penury and oblivion. Fate at least would not be vindictive.

Labé would be allowed to return to sea and attempt to rebuild his modest fortune, until plagued by his old wounds he was forced to give up. One day, when the new knowledge had begun to spread and both the causes of scurvy and the remedies to be applied were better understood, he would be able to look back on it all and be philosophical about the whole tragic and absurd business.

Villefeix would reappear quietly, when all the talk had died down. His dreams of wealth in Peruvian gold mines had come to naught. He had suffered hardships, felt cold and hunger, and risked his life in scrapes which he took care never to discuss. His career as a priest was not at an end, although the chair once reserved for him in St James's Convent in Paris belonged to another life. He went back to his native Périgord and helped his brother with his parish duties. Did he ever look back to that nightmarish voyage and realise that he had been the first to bring the strange ritual of Christianity to a dark, pagan land? We may imagine that he did, some night when the wind howling outside the haven of the country presbytery reminded him of the gale that screamed through the rigging of the *St Jean Baptiste* in faraway Lauriston Bay.

If ever he did, he kept his peace — or if he was moved, one day, to set down his adventures on paper, his account has been lost.

And are there today still, in Peru or Chile, some descendants of young de Surville or of Charenton the second-lieutenant, both of whom had deserted in Callao? Did they settle in South America? Or die there in some brawl, or at the hands of Indians, or of fever in a hut somewhere in some nameless village? Or did they creep back to France eventually, in homesickness — twenty years later maybe, when the great Revolution sweeping through France obliterated so many records and shortened so many memories?

Fate had long since dropped the puppets' strings and forgotten them. The mystery of Davis Land she had used as a lure was never solved. Like other imaginary islands it vanished together with the trappings of Eldorado with which man's imagination had endowed it. The colony of Jews was never found. But no great British fleet ever lay at anchor in some rich Pacific harbour to control the Ocean and keep out France, as de Surville had feared. The Pacific was large enough for them both and, anyhow, the centuries-long pattern of enmity between the two countries was slowly to disappear.

The dreams and sufferings of men faded away gradually into the great picture of man's endless dreaming and suffering. And the dreamer sleeps on in his unknown grave.

INDEX

Acapulco, 118
Amboina, 103
Aumezat, M., 17
Austrialia del Espiritu Santo, 71

Babuyans, 47
Bashi Islands,43ff, 108
Batan Islands, 82, 91
Batavia, 22, 103
Bay of Islands, 93
Bejeador, Cape, 47
Bellavista, 117
Bengal,12-14, 23, 25-6, 111, 112
Bougainville, Louis-Antoine de, 16, 99
British East India Company, 21, 24ff
Byron, John, 16

Cagnard, François, 68
Calcutta,12-13, 21, 25
Callao, 103, 113, 116, 118ff
Canton, 22, 25, 31, 41
Cape Horn, 123
Cape Maria van Dieman, 76, 79
Cape Reinga, 75
Carvalhos, 37
Chandernagore, 9ff, 21ff, 86
Chevalier Cove, 93
Chevalier, Jean-Baptiste, 13ff, 21ff, 26
Chilca, 113
Chile, 111
China, 22, 23, 30, 47
Choiseul Island, 58
Clive, Robert, 10
Concorde, La, 17
Cook, James, 5, 24, 68, 72, 76, 79, 81-2, 99
Coral Sea, 72

Dampier, William, 47, 109
David, Captain (*see* Davis), 109
Davis Land, 109-11, 120, 124
Davis, Edward, 109-10
de St Paul, Captain, 32-3, 50, 83
de Surville, Elisabeth, 92
de Surville, Hughes, 27, 76, 83, 88, 102,120-1
de Surville, Jean-François-Marie, 5, 14ff, 21ff, 33ff, 44ff, 67ff, 81ff, 91ff, 102ff, 118
de Surville, Marie, 123
de Surville, René, 17
Delessert brothers, 14, 22
Deliverance, Islands of, 75
Dolphin, the, 25
Doubtless Bay, 82ff, 102
Dulucq, Pierre, 27, 54, 67, 76, 80, 83, 92, 101, 102, 112-13
Dupleix, François, 10-12
Dutch East Indies, 103

Easter Island, 109, 120
Egan, Demetrius, 116, 118-9
Endeavour, the, 24, 68, 76, 79

Fiji, 72
French India Company, 16, 18

Goat Island, 47
Guam, 108

Hervé, Julien, 68
Hokianga Harbour, 75

Juan Fernandez Islands, 111

Kasimbazar, 18, 23
King George III's Island (*see* Tahiti), 23

Labé, Guillaume, 5, 7ff, 21ff, 32ff, 44ff, 53ff, 68ff, 81, 82ff, 91ff, 102ff, 116ff
Lauriston Bay, 81, 105, 123
Law de Lauriston, Jean, 12, 15, 17, 25, 29-30, 81
Law, Baron John, 18

Le Roux, Gabriel, 33
Leroy, Pierre, 113
Lind, James, 68
Lopez, Juan, 112
Lova, 105, 120
Luzon, 47, 53

Madras, 37
Mahrattas, 28-9
Malacca, Straits of, 22, 32, 34
Malaita Islands, 68
Mangin, 29
Manila, 25, 103
Manning Strait, 58
Maori, 76, 80ff
Marsden, Samuel, 93
Mascarin, le, 13, 17, 20
Masulipatnam, 23, 29
Maurepas, M., 55
Mexico, 105
Monneron, Pierre, 5, 14, 26, 47ff, 54, 69, 76, 81, 83, 92, 95, 102, 104, 117ff

New Caledonia, 72
New Guinea, 22, 103, 120
New Hebrides, 71-25
New Zealand, 22, 53, 75, 107, 120
Nicobars, 33-4
Ninety Mile Beach, 76
Noble Savage, 48
North Cape, 79

Panon, Jean-Jacques, 28
Papua, 59-66
Pelew Islands, 55
Peru, 104, 110, 113-4, 118
Philippines, 22, 55, 103
Pondicherry, 9ff, 29, 30-31, 104, 117
Port Praslin, 59-67, 68ff, 105, 120
Port-Louis, 92-3, 123
Pottier de l'Horme, Jean, 5, 27, 41, 47ff, 54, 65, 69, 72, 74, 76, 92, 97, 101, 102-5, 108, 111, 112

Pulo Condore, 43
Pulo Tioman, 37
Pulo Varella, 35

Quiros, 71

Ranginui, 101, 105, 106, 111
Refuge Cove, 97

Santa Isabel, 58, 68
Shah Alam, 18
Solomon Islands, 5, 75, 80
South America, 103, 108, 110
Southern Continent, 25-6
Spain, 104, 119, 120
St Jean Baptiste, 14, 19, 22, 24ff, 32ff, 43ff, 49, 58, 67, 71, 72, 76ff, 80ff, 92ff, 102, 104, 107, 110ff, 116ff
Surville Cliffs, 79, 80

Tahiti, 25, 26, 76, 99
Taillefer, 112
Tasman, Abel, 5, 75, 76, 86
Tasmania, 53
Three Kings Islands, 76
Timor, 103
Tonga, 72
Torres Strait, 103
Trengannu, 39-40

Ulawa, 69ff

Villefeix, Father Paul-Antoine de, 27, 73, 76, 81, 91-4, 102, 123

Wallis, Captain Samuel, 25, 76

Yanaon, 23, 26, 27-8

Selected Other Publications by John Dunmore

French Explorers in the Pacific. Vol 1: The Eighteenth Century
French Explorers in the Pacific. Vol 2: The Nineteenth Century
Monsieur Baret: First Woman Around the World 1766–68
The Pacific Journal of Louis-Antoine de Bougainville, 1767–1768
The Journal of Jean-François de Galaup de la Pérouse, 1785–1788, vol. 1
The Journal of Jean-François de Galaup de la Pérouse, 1785–1788, vol. 2
Le Voyage de Lapérouse (2 volumes)
La Peyrouse dans l'Isle de Tahiti
Where Fate Beckons: The Life of Jean-François de la Pérouse
Storms and Dreams. Louis de Bougainville: soldier, explorer, statesman.
Pacific Explorer: The Life of Jean-François de la Pérouse, 1741–1788
La Pérouse : Explorateur du Pacifique
The Expedition of the St. Jean-Baptiste to the Pacific, 1769–1770: from the Journals of Jean de Surville and Guillaume Labé
Chronology of Pacific History
Visions & Realities: France in the Pacific, 1695–1995
Who's Who in Pacific Navigation
Mrs Cook's Book of Recipes for Mariners in Distant Seas

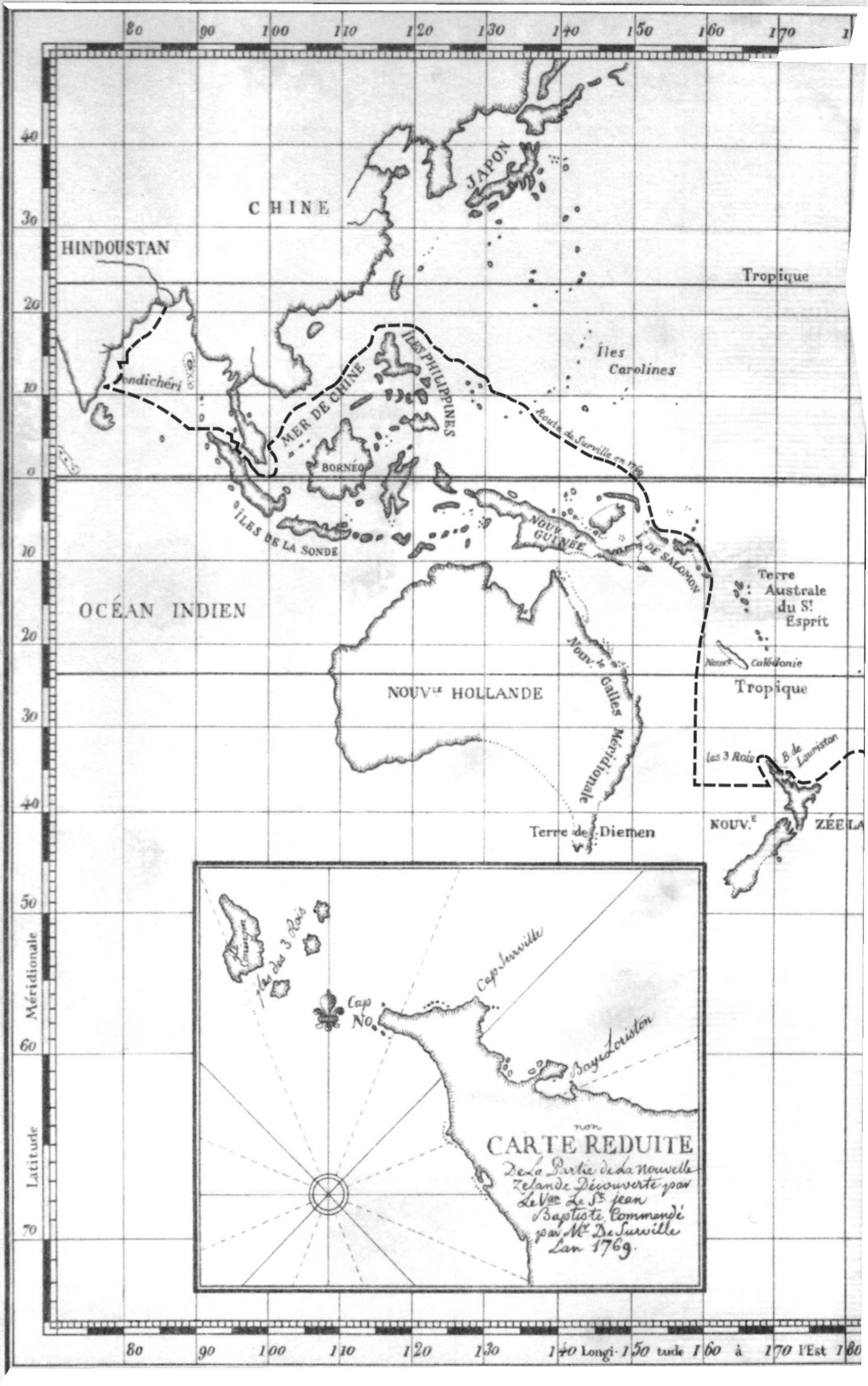

CHINE
JAPON
HINDOUSTAN
Tropique
Îles Carolines
MER DE CHINE
ÎLES PHILIPPINES
ondichéri
Route de Surville en 1769
BORNÉO
ÎLES DE LA SONDE
NOUV.LE GUINÉE
DE SALOMON
Terre Australe du S.t Esprit
Nouv.le Calédonie
Tropique
OCÉAN INDIEN
NOUV.LE HOLLANDE
Nouv.le Galles Méridionale
les 3 Rois
B. de Lauriston
Terre de Diemen
NOUV.E ZÉELA
Cap Surville
Isles des 3 Rois
Cap No.
Baye Loriston
non
CARTE REDUITE
De La Partie de La nouvelle Zelande Découverte par Le Vau Le S.t jean Baptiste Commendé par M.r De Surville Lan 1769.
Longi- tude à l'Est
Latitude Méridionale